THE ADOPTION OF SUSTAINABLE AGRICULTURAL TECHNOLOGIES

For my mother, Adélia.

The Adoption of Sustainable Agricultural Technologies

A Case Study in the State of Espírito Santo, Brazil

Hildo M. de Souza Filho
Department of Economics
Federal University of Espírito Santo
Brazil

LONDON AND NEW YORK

First published 1997 by Ashgate Publishing

Reissued 2018 by Routledge
2 Park Square, Milton Park, Abingdon, Oxon, OX14 4RN
52 Vanderbilt Avenue, New York, NY 10017

Routledge is an imprint of the Taylor & Francis Group, an informa business

A Library of Congress record exists under LC control number: 97074458

ISBN 13: 978-1-138-38443-9 (hbk)
ISBN 13: 978-1-138-38444-6 (pbk)
ISBN 13: 978-0-429-42754-1 (ebk)

Contents

Figures and tables

Preface

The effects of agriculture on the environment have become a worldwide concern. Although in many countries the introduction of mechanical and chemical technologies has boosted agricultural production, this positive outcome has been offset by negative side-effects. After several years of massive incentives to promote production in agriculture, governments are now facing the problems of loss of arable land, loss of water quality, deforestation, desertification and loss of genetic resources. In some developing countries, social and economic impacts have also been disastrous. Despite the achievement of positive rates of growth in aggregate rural income, poverty has been exacerbated by increasing inequalities in land and income distribution, and in regional development. Several attempts have been made to connect issues such as economic growth, exploitation of natural resources, conservation, quality of life, income distribution and poverty. However, only after the dissemination of the concept of sustainable development was the joint treatment of these themes effectively considered in economic development thought. There is a growing consensus that environmental conservation and better standards of living must be pursued simultaneously.

Agricultural technology has played an important role in shifting production systems towards greater sustainability. Cleaner technology options such as fine-tuned conventional techniques and new production systems can simultaneously deliver environmental conservation and fairer social systems. However, only recently has the diffusion of these practices become significant. Non-economic reasons such as environmental concerns, health concerns, and personal beliefs, have been pointed out as the leading forces behind farmers' adoption behaviour. However, economic considerations such as profitability of the innovations and access to capital sources can also play fundamental roles. Many of these factors are affected by elements like market prices or government prices which are beyond farmers' control and which can change over time, between countries and also between regions.

This study is an attempt to provide an understanding of the determinants of adoption and diffusion of sustainable agricultural technologies in the State of Espírito Santo, Brazil. Economic, social and environmental aspects of the Green Revolution in the country and particularly in that state are examined as a means of establishing the background. A survey conducted in 1994 provided data on 141 farms. This information is analysed using two econometric techniques: logit/probit models and duration analysis. The role of farms' and farmers' characteristics and variables that are external to them (and change over time) are examined. The results highlight the importance of contact with non-governmental organizations, integration with farmers' organizations, health and environmental concerns, availability of family labour, farm size, and physical characteristics of farms. However, economic variables (change in relative prices), which affected agricultural profitability and were beyond farmers' control, accelerated or decelerated the diffusion process. Specifically, the diffusion of sustainable agricultural practices was eased by farmers' economic constraints. On one hand, the decline in output prices squeezed agricultural profit and many farmers faced difficulties in buying external inputs. On the other hand, labour became cheap due to the economic crisis, which made low-external-input practices a more attractive option for family small-holdings.

Acknowledgements

Writing a book like this would not be possible without the help, support and encouragement of a number of people and institutions. First I must thank Dr Trevor Young for his expertise, meticulous advice and support throughout, and Dr Michael Burton for promptly offering guidance and assistance. I have also to thank a fellow researcher, Mr Francesco Lissoni, who made decisive suggestions at an early stage of this research. In Brazil, I am indebted to Mr Adilson Vivas of the Secretary of Agriculture of Espírito Santo, and the members of the APTA (Alternative Technology Programmes Association) and affiliated organizations in Espírito Santo, particularly Mr Eduardo Soares and Mr Klaus Nowotny, who have aided me in the field survey. I owe an especial debt of gratitude to Mrs Jane Cabot for reading and correcting the drafts. My thanks are also due to my wife for love and encouragement, and to my sons, Daniel and Mateus, for lighting up my days in Manchester. My research was financially supported by the Universidade Federal do Espírito Santo, and the Conselho Nacional de Desenvolvimento Científico e Tecnológico (CNPq). The field research in Brazil was also financially supported by the WWF - Fundo Mundial para a Natureza. The responsibility for what follows remains entirely mine.

List of abbreviations

ASSESSOAR	Studies, Orientation and Assistance Association
CAMP	Multi-professional Advising Centre
AGAPAN	*Gaucha* Association for Natural Environment Protection
AGF	Federal Government Purchase
ANFAVEA	Automobile Makers National Association
APPRLT	Laranja da Terra Small Farmers' Association
APPRSP	Serra Pelada Small Farmers' Association
APSAD-VIDA	Santa Maria Farmers Association for Life Protection
APTA	Alternative Technology Programmes Association
AS-PTA	Alternative Agriculture Project Consultants
CAPA	Small Farmers' Advice Centre
CENARGEN	National Genetic Resource and Biotechnology Research Centre
CEPLAC	Executive Commission for the Recovery of Cocoa Farming
CETAP	Centre for People's Alternative Technologies
CIER	Rural Education Integrated Centre
CNPMA	National Research Centre for Monitoring and Assessment of Environmental Impact
CNPSo	National Soybean Research Centre
CNPT	National Wheat Research Centre
CTA	Oricuri Alternative Technologies Centre
CTC	Cotrijuí Training Centre
DIEESE	Inter-Union Department of Statistics and Socio-Economic Studies
DNOCS	National Department of Constructions Against Droughts

EGF	Federal Government Loan
EMBRAPA	Brazilian Agricultural Research Corporation
FGV	Getúlio Vargas Foundation
GDP	Gross Domestic Product
IAA	Sugar-cane and Alcohol Institute
IBGE	Brazilian Institute of Geography and Statistics
IGP-DI	General Price Index - Domestic Availability
INCRA	National Institute for Colonization and Agrarian Reform
INPA	National Institute for Amazon Research
IPM	Integrated Pest Management
LEISA	Low-External-Input and Sustainable Agriculture
MEPS	Educational and Promotional Movement of Espírito Santo
MOC	Feira de Santana Community Organization Movement
NGO	Non-Governmental Organization
PATAC	Community Adapted Technologies Application Programme
PDRI	Integrated Rural Development Programmes
PGPM	Minimum Price Guarantee Policy
PIN	National Integration Programme
PNAD	Household Sample National Research
POLOCENTRO	Savanna Development Programme
POLONORDESTE	Northeast Integrated Areas Development Programme
PPI	Multiyear Irrigation Programme
PROCACAU	Programme for Cocoa Plantations Recovering
PROTERRA	National Programme for Land Redistribution and Incentive to the North and Northeast Agro-industry
PROVARZEAS	National Programme of Wetland Drainage
PTD	Participatory Technology Development
R&D	Research and Development
SPVEA	Superintendence of the Plan for Economic Valorization of the Amazon
SUDAM	Amazon Development Superintendency
SUDENE	Northeast Development Superintendency
TAPS	Brazilian Association of Alternative Technology for Health Promotion
WCED	World Commission on Environment and Development

1 Introduction

The effects of agriculture on the environment have become a worldwide concern. Although in many countries the introduction of mechanical and chemical technologies has boosted agricultural production, this positive outcome has been offset by negative side-effects. After several years of massive incentives to promote production in agriculture, governments are now facing problems such as the loss of arable land, loss of water quality, deforestation, desertification, and loss of genetic resources. In some developing countries, social and economic impacts have also been disastrous. Despite the achievement of positive rates of growth in aggregate rural income, poverty has been exacerbated by increasing inequalities in land and income distribution, and in regional development. The Brazilian Green Revolution is an example of these imbalances. In Brazil, subsidized rural credit induced the adoption of technological patterns that generated a process of rapid increase in agricultural production along with environmental degradation and social disruption.

It is against this background that *sustainable development* has emerged as a theme in economic development. The central concern within the literature on sustainable development is to maintain welfare over time. This is expressed in the much quoted WCED's definition:

> Sustainable development is development that meets the needs of the present without compromising the ability of future generations to meet their own needs (WECD, 1987, p.43).

In order to meet the needs of the present generation, developing countries must improve, rather than just maintain, the welfare of a significant part of their population. For them, sustainability implies that both better standards of living and environmental conservation should be pursued. Here, economic growth, exploitation of natural resources, conservation, quality of life, income distribution and poverty form a vector of connected problems which have to be balanced in order to achieve sustainable development.

Agricultural technology has an important role to play in shifting production systems towards greater sustainability. Cleaner technologies, which have been classified in the literature as sustainable, can simultaneously deliver environmental conservation and fairer socioeconomic systems. Sustainable agricultural technologies may take a number of forms. Some refer to specific practices or systems (e.g. organic agriculture, low tillage, and integrated pest management), while others have a broader meaning (e.g. alternative agriculture and LEISA - low-external-input and sustainable agriculture). Most of these forms of production are currently well known in agricultural economics and rural sociology, and the socioeconomic assessment of sustainable agricultural technologies is an increasing area of research.

This study is an attempt to provide an understanding of the determinants of adoption and diffusion of sustainable agricultural technologies in the State of Espírito Santo, Brazil. In the early 1970s, the agriculture of the State was characterized by the prevalence of family smallholdings and by a more even distribution of land than other States of Brazil. During the second half of that decade, however, the diffusion of chemical and mechanical innovations (mainly promoted by inadequate governmental policies) led to changes in agrarian structure. Although agricultural production increased, rural-urban migration accelerated due to the introduction of labour-saving technologies. Redundant labour entered low-paid urban jobs and thereby aggravated the social crisis in the cities. The environment was negatively affected as problems related to deforestation, soil depletion, water pollution and human poisoning became a threat to farmers, rural workers, local communities and agricultural production. In the early 1980s, few farmers in Espírito Santo were using sustainable techniques. By the middle of that decade, however, the number of farmers changing to practices such as organic fertilization and cheap, environment-friendly forms of plant protection, started to grow. During this period, the Brazilian economy suffered severe crisis and the agricultural sector was affected by changes in governmental policies, output and input prices.

Several non-economic reasons such as environmental concerns, health concerns and personal beliefs, have been suggested as the leading forces behind farmers' behaviour towards adoption of sustainable technologies. However, economic considerations such as the profitability of the innovation and access to capital sources can also play fundamental roles. Many of these factors are affected by elements (e.g. market prices and government policies) that are beyond farmers' control and which can change over time, between countries and even between regions. It is suggestive that the number of adopters of cheap, environmental-friendly technologies in Espírito Santo's agriculture started to grow in a period of significant changes in the economic environment. Thus, the investigation here will consider not only the role played by non-economic aspects, but also the effect of economic factors on the process of adoption and diffusion.

This book is organised into nine chapters. Chapters 2 and 3 give background information on economic, social and environmental aspects of the Green Revolution in Brazil and Espírito Santo. Chapter 2 gives a historical and regional view of Brazilian agriculture. An overview of rural welfare indicators and the effects of the Green Revolution on the environment and human health is also provided. The chapter ends with a description of recent efforts of governmental and non-governmental organizations towards sustainability. Although the changes in agriculture have followed the same direction as all other Brazilian States, local characteristics have affected the speed and intensity of diffusion of Green Revolution technologies. Thus, chapter 3 moves on to the context of Espírito Santo in order to provide the background to the empirical analysis of chapter 8.

An overview of several issues related to the concept of sustainable development is given in chapter 4. Some theoretical approaches to the theme are reviewed, including different views on sustainable agriculture. Agricultural technologies that are considered in the literature to be sustainable are identified. The aim of the chapter is to define concepts such as sustainability, sustainable agriculture, and sustainable technologies.

The main themes in the theory of adoption and diffusion of technologies are explored in chapter 5. The major objective there is to provide a background to the methodological approach of our empirical investigation. The chapter brings a review of theories of technological change and models of individual decision-making behaviour. There is no discrimination in favour of the literature on agricultural sector modelling. This broader examination allowed us to explore methodological approaches such as duration analysis which have not been used to investigate adoption in agriculture but which appear promising.

Adoption of sustainable agricultural technologies is a new area of study in rural sociology and economics. Empirical analyses are not numerous, as most sustainable practices are in the early stage of diffusion. The objective of chapter 6 is to review the growing literature concerned with this theme. It presents a discussion of economic and non-economic reasons to adopt, an examination of the existing barriers to widespread adoption and highlights the economic appeal of sustainable systems. The role played by information, farms' and farmers' characteristics, and government policies are considered. This chapter provides the basis for formulating hypotheses about the determinants of adoption of sustainable agricultural technologies.

Two econometric approaches are reviewed in chapter 7. Probit/logit models have been widely used to test hypotheses about technology adoption, while duration analysis has only recently been applied for this purpose. The main objective of this chapter is to depict the econometric basis of the empirical analysis reported in chapter 8, where the determinants of the adoption of sustainable agricultural technologies in Espírito Santo are investigated. A survey conducted in 1994 provided data on 141 farms in the State. This information is analysed together with data from secondary sources.

Finally, chapter 9 presents the concluding comments of the book.

2 Economic, social, and environmental aspects of the Green Revolution in Brazil

Introduction

The social and economic impact of the Green Revolution in the Third World has been very distinct between countries. Several factors explain these differences, but the political background of the societies in which the technological packages were introduced is decisive. In Brazil, a process of rapid diffusion of Green Revolution innovations started in the second half of the 1960s. In this period, the military government promoted decisive economic restructuring, which allowed it to reorient agricultural policies. Subsidized rural credit induced adoption of predetermined technological patterns and generated a process known as *conservative modernization*. They were implemented over a highly concentrated agrarian structure and a social system that privileged the elites. The social effects of this process were so profound that any attempt at examining sustainability in Brazilian agriculture cannot ignore them.

Economic, social, and environmental aspects of the Brazilian Green Revolution are examined in this chapter. This broader view will allow a qualitative assessment of whether it could be characterized as an example of agricultural (un)sustainable development. A brief description of the historical evolution of the country's agriculture since the Second World War is provided. Given Brazil's immense territory and marked regional differences, a regional perspective of Brazilian agriculture is given. Some rural welfare indicators and the effects of the Green Revolution on the environment and human health are examined. Finally, the chapter offers an account of governmental and non-governmental efforts towards sustainable agriculture.

The historical phases of the Green Revolution in Brazil

According to Martine (1989) the Brazilian Green Revolution comprises three distinct phases: the post-war to 1965, 1965 to 1979, and after 1979. During the first period, the agricultural input industry was not completely established in the country; machinery and other inputs were largely imported. Technological diffusion was slow and did not reach many parts of the territory. In fact, it took place predominantly in the most industrialized regions of the South and South-East. A process of more widespread diffusion would only occur later.

The second period, 1965 to 1979, was characterized by deep changes in the countryside. The State supported agricultural activities by providing a combination of large quantities of subsidized credit, special programs, and research and extension services, which allowed the introduction of the Green Revolution technological packages. This phase was also characterized by the establishment of the agro-industrial sector (processor and input industries). Most large farmers, mainly those who were located on better land, were capitalized and adopted the new technologies. Even groups of smallholdings, which were integrated to agro-industries, co-operatives, and international markets, changed their production methods. However, many peasants and non-capitalized small farmers, who farmed subsistence and local market crops, did not get access to the new technologies. Most were located in poor regions and were not covered by government policies. For most export crops, physical yields increased dramatically during the period, while the performance of local market products declined.[1]

At the end of the 1970s, the diffusion of Green Revolution technologies started to slow. This was an important characteristic of the third phase. The financial crisis and the adjustment policies adopted by the federal government affected credit sources and, consequently, the provision of subsidized rural credit, the most important agricultural policy to date. However, export and import-substitution activities, both related with external adjustment policy, continued receiving significant amount of credit. A policy of energy substitution was actively pursued, such as the Proalcohol Programme. After 1983, as an alternative for credit constraints, the federal government intensified the Minimum Price Guarantee Policy (PGPM).

A regional perspective

Brazilian agriculture has been characterized by extreme spatial heterogeneity. In most industrialized areas of the South and South-East Regions, the Green

Revolution technologies have reached higher levels of diffusion, while in many parts of the North and North-East erosive forms of low-input agriculture, such as slash-and-burn, still remains (see Tables 2.1 and 2.2; and map of Brazil in Figure 2.1). Thus, the lower level of diffusion of chemical inputs in the North and North-East, as showed in Table 2.1, does not mean that farmers in these regions have adopted alternative practices, such as organic fertilization. As will be seen in this chapter, environmental problems related to agriculture in Brazil come from both Green Revolution technologies and traditional, destructive farming systems. A brief description of the Brazilian Region's agricultural changes, particularly during the 1960s and 1970s, is presented here to indicate the regional differences that currently characterize the country's agriculture.[2]

The North Region

The North Region comprises the States of Acre, Rondônia, Amazonas, Roraima, Pará, and Amapá. At the end of the nineteenth century, the region was the world's largest supplier of rubber. This was produced in a costly semi-servile system, known as *aviamento*, in which the raw material was obtained by the rubber tapper from native trees and traded through a complicated network of power relations and commercial interests. When the British introduced rubber plantations in Asia, the Amazon production collapsed. This decline was followed by an increase in nut gathering, which allowed for the surviving of the *aviamento* socio-structure. With the exception of these activities, up to the early 1940s, the region's agriculture was fundamentally characterized by subsistence production. After World War II, Japanese immigrants started to farm black pepper and jute in the State of Pará. Jute production followed the traditional *aviamento* system, while a cooperative system was developed to support black pepper production.

Development policies for the Amazon Region started to emerge during the1950s, when two institutions were created: the SPVEA (Superintendence of the Plan for Economic Valorization of the Amazon) and the INPA (National Institute for Amazon Research). The objective of the SPVEA was to integrate the region with the rest of the nation. Its most important project was the Belém-Brasília motorway, which allowed new settlement by migrants. The INPA's initial projects were related to botany, but it began to deal with agronomic research in the 1970s.

During the 1960s and 1970s, several programmes, plans, and incentives were created to stimulate economic development in the Amazon area. The military government's concern with defence and national integration led to the 1970 PIN (National Integration Programme), which set up major, and largely incomplete projects to build giant motorways (Transamazônica and Cuiabá-Santarém).

Table 2.1
Use of chemical inputs in Brazilian agriculture: 1960-85

Regions	Farms using fertilizers (%)					Farms using plant protectors (%)		
	1960	1970	1975	1980	1985	1975	1980	1985
North	1.7	1.9	2.4	5.5	2.4	12.0	19.5	18.3
North-East	4.8	6.7	7.7	13.1	7.0	33.8	46.2	40.5
South-East (-SP)	16.6	27.6	36.1	52.8	47.0	72.1	80.3	73.8
São Paulo	26.6	47.5	62.0	77.7	70.0	74.6	83.4	78.9
South	24.8	33.8	42.8	60.9	56.0	78.3	85.4	82.5
Centre-West	0.9	4.2	14.1	28.9	31.5	61.8	73.5	72.6
Brazil	13.2	18.6	22.3	32.1	26.0	51.0	60.3	54.9

Sources: Kageyama (1986) and IBGE (1985)

Table 2.2
Availability of tractors in the Brazilian agriculture

Hectares of farmed land / tractor: 1960-85

Regions	1960	1970	1975	1980	1985
North	45,976	16,833	16,198	5,673	6,563
North-East	14,626	7,516	3,924	1,790	2,206
South-East (-SP)	5,598	2,827	1,536	748	668
São Paulo	641	277	186	134	127
South	1,563	603	276	180	167
Centre-West	24,353	6,899	2,866	1,562	1,083
Brazil	3,407	1,483	855	572	562

Sources: Kageyama (1986) and IBGE (1985)

Figure 2.1 Map of Brazil: regions and States

In the early 1980s there were official attempts to colonize the area with smallholding settlements, but the result was the establishment of large farms for cattle grazing. Industrial groups and large farmers from the South showed great interest in the benefits offered by the SUDAM (Amazon Development Superintendency) and bought huge quantities of land in the area.[3]

Crop plantations, such as cocoa, sugar-cane, black pepper, and rubber trees were also stimulated by governmental benefits during the period. This colonization process brought serious social problems to the region. As the condition of receiving the fiscal incentives was an effective occupancy of land, in some instances enterpreneurs violently removed Indians and peasants who lived without property rights in remote areas.

The immigration flow during the 1970s and 1980s was intense. Most immigrants were motivated by job offers on cattle grazing farms, deforestation works, motorway building, wood industry, and by settlement programmes. The infrastructure however was precarious. A large amount of agricultural production was lost because of lack of transport and storage facilities. At the end of the 1970s, the idea of transforming the Amazon Region into an area of surplus agricultural production was abandoned. The big motorway projects were deactivated and the government set up policies to restrain new and monitor old agricultural projects in ecologically sensitive areas. More emphasis is now given to problems related to agrarian conflicts and to regularization of property rights.

The Centre-West Region

Until the late 1950s, most of the Centre-West territory (States of Mato Grosso, Mato Grosso do Sul, Goiás, and Tocantins) was free from agricultural activities. Only in the 1960s, particularly after the building of Brasília and the network of motorways that link the city with the rest of the country, were agricultural activities expanded in the region. The north of the States of Tocantins and Mato Grosso border the Amazon Region and received the benefits offered by the programmes created for that area. The States of Goiás and Mato Grosso do Sul are closer to the South-East, the most developed region of the country. Consequently, they became an area of agricultural expansion (mainly with cattle grazing activities) when land became relatively more expensive.

The region's most important development programme was the POLOCENTRO (Savana Development Programme), whose main objective was to incorporate 3.7 million hectares to the agricultural land of the States of Minas Gerais, Goiás, and Mato Grosso. Given the area's special characteristics of soil fertility, acidity and topography, the government had an important role in terms of technology development and financial support for agriculture. The POLOCENTRO was conceived on the basis of subsidised and oriented credit system, in which the official extension service, with the participation of the private sector (especially in services related to mechanization), took part in several phases of the agricultural production.

In general, the Centre-West is characterized by a predominance of large farms, whose main activities are extensive cattle grazing and export commodity plantations such as soybeans, coffee and cotton, although production of traditional crops, such as maize and rice, has recently increased. Government programmes have promoted these large enterprises, and also some of their *by-products*: speculation in the land market and land property concentration.

The North-East Region

The North-East comprises nine States: Maranhão, Piauí, Ceará, Rio Grande do Norte, Paraíba, Pernambuco, Alagoas, Sergipe, and Bahia. It is economically, socially and physically diversified. Although the region has been the focus of the greatest number of the country's agricultural development plans and programmes, it presents the most backward socio-economic agricultural systems. Land property is extremely concentrated and government actions have mainly favoured large landowners. The effect of recurrent droughts and the political dominance of mercantile interests have contributed to the precarious condition of the region's agriculture. Despite constant efforts, very limited positive results have been achieved in terms of agricultural restructuring. The failure has been attributed to the rigidity of the region's agrarian structure, which has blocked the diffusion of development benefits.

The SUDENE (North-East Development Superintendency), the region's most important agency of development, was created in 1959. During the 1960s, because of the complex network of political and economic interests, the SUDENE had difficulties in imposing its programmes on other government institutions such as the DNOCS (National Department of Constructions Against Droughts), the IAA (Sugar-cane and Alcohol Institute), the CEPLAC (Executive Commision for the Recovery of Cocoa Farming), the INCRA (National Institute for Colonization and Agrarian Reform), the Bank of Brazil, and the North-Eastern Bank, which also operated in the Region.

In the 1970s, regional policies were progressively centralized by the federal government in order to attain the objectives of the National Development Plan. Four large programmes affected the North-East during this period: the PIN (National Programme of Integration), the PROTERRA (National Programme for Land Redistribution and Incentive to the North and North-East Agro-industry), the POLONORDESTE (North-East Integrated Areas Development Programme), and the *Sertanejo* Project (the North-East Semi-Arid Region Special Programme Support). The recommendations of the PIN for the North-East were incorporated into the PPI (Multi-Year Irrigation Programme), whose main objective was to secure employment and income for the population living in the drought areas. Paradoxically, the Programme was subsequently criticized for its negative social effects. Projects were selected on the basis of their economic pay-off and maximum commercial production. Many small farmers did not receive any benefit, while the land of others was unfairly expropriated. The objectives of the PROTERRA were to improve land distribution through expropriation of large farms, and provide subsidized credit to agricultural production. By the end of the 1970s, the impact of the programme over the North-East agrarian structure had been very limited and the PROTERRA had provided financial support for the adoption of Green Revolution technologies only by medium and large farmers. The POLONORDESTE, through the PDRI (Integrated Rural Development Programmes), gave support to settlement, irrigation and urban development projects, and contributed to the improvement of the region's infrastructure. However, like the other programmes, it did not change the backwardness of the region. The Sertanejo Project suffered the same fate.

On the whole, the North-East agriculture is one of low productivity. In terms of production, the region can be divided into four different areas: the States of Maranhão and Piauí, which border the Amazon Forest and where, still in the 1960s, traditional agriculture production systems (crops and livestock) were predominant; the States of Ceará and Rio Grande do Norte, in which these traditional systems exist along with natural fibres production (cotton and sisal); the more technologically intensive sugar-cane plantations, which are located next to the coastal areas of the States of Paraíba, Pernambuco, Alagoas, and Sergipe; and the State of Bahia, which presents diversified agricultural production (cocoa, coffee, tobacco, cassava, beans, milk and beef cattle grazing).

The South-East Region

The South-East comprises four States: Minas Gerais, Rio de Janeiro, Espírito Santo, and São Paulo. Its economic formation is related to the history of Brazilian coffee and the country's process of industrialization. There are large differences between the States in terms of agricultural production and agrarian structure. Levels of technological intensification in São Paulo, for instance, are the highest in the country. It contrasts with poor areas of the Serra do Mar in the States of Rio de Janeiro, Espírito Santo and Minas Gerais, where coffee plantations flourished during the second half of the last century. After many years of constant exploitation, these areas lost their soil's natural fertility;

extensive agriculture became economically unfeasible and, with few exceptions, mechanization is restrained due to hilly topography.

The State of Minas Gerais has distinct agricultural sub-regions: the Jequitinhonha Valey, which due to its proximity to the North-East Region was included in the programmes that benefit the Brazilian semi-arid zone; the Mineiro Triangle, which has levels of technological intensification similar to its neighbour São Paulo; and the Mata Zone, which has traditonally been devoted to coffee and milk production. During the 1970s, areas with intensive coffee farming increased in the Mineiro Triangle, new breeds for milk production were introduced in the Serra do Mar (Mata Zone), and savanna areas were brought into agricultural production (particularly with large plantations of soybeans and wheat).

The golden age of Rio de Janeiro's agriculture ended last century with the demise of the slavery coffee plantation system. Although farming of traditional crops has increased since then, impoverished soils and difficulties with mechanization have created barriers to agricultural development, mainly in the hilly areas of the Serra do Mar. In the north of the State, sugar-cane plantations have received extensive benefits from the IAA (Sugar-cane and Alcohol Institute) and the Proalcohol Programme.

In the State of São Paulo is located the heart of the Brazilian industrial and agricultural sectors. Although the Green Revolution has occurred in several regions of the country, for historical reasons the highest levels of intensification were achieved in São Paulo. During the 1960s and 1970s, the area with potatoes, oranges, tomatoes, sugar-cane, soybeans and cotton (products for which technological packages were first developed) increased, while both traditional crops (rice and beans) and products whose technological development was at an early stage (banana, onions, maize, peanuts, cassava, coffee and tea) decreased. Several factors contributed to this restructuring: international prices, agro-industrialization and subsidized rural credit policy, which favoured the technological packages.

The landmark in the recent history of Espírito Santo's agriculture was the 1960s' coffee eradication programmes. The economy of the State was highly dependent on coffee production and these programmes caused social and economic problems. The government tried to diversify the economy by promoting the establishment of agro-industries, the expansion of cattle grazing activities, and reafforestation. This policy however aggravated rural-urban migration problems. By the middle of the 1970s, when coffee farming was resumed, the level of technological intensification of the State's agriculture rapidly increased (chapter 3).

Because the South-East was the most developed region of the country, it did not receive the enormous financial resources that characterized the large projects of the National Plans of Development. However, the benefits of the general class of policies that supported the Green Revolution, such as subsidized rural credit, subsidized input prices, research and extension, were disproportionately distributed in favour of the region. Economic policy, in a broader sense, tended to favour the most developed areas of the Country and, consequently, their agriculture.

The South Region

The common characteristic of the three States of the South Region (Paraná, Santa Catarina, and Rio Grande do Sul) is the predominance of medium and small farms, whose historical formation is related to the eighteen century immigration policies. During this period, new settlements of European immigrants were established by the government to occupy the region. Apart from subsistence production, they developed activities to supply the other Brazilian regions with livestock, leather, jerked beef, and dairy products. Some extractive activities (wood, charcoal, and Paraguay tea) and activities based on European immigrant traditions (wine) were also carried out. Agricultural production grew extensively up to the early 1960s, when scarcity of new land and governmental policies fostered a process of intensification.

The Green Revolution in the region was not only supported by the national policy of subsidized rural credit but also by specific schemes. Agricultural cooperatives have been specially important in the South in terms of services provision (mechanization, credit distribution, etc.) and commercialization. Food processing industries, mainly the soya complex in Paraná and in Rio Grande do Sul, and pork and chicken in Santa Catarina, have also influenced the process of agricultural intensification. A protected market, institutional research and the government's extension service helped to promote wheat production. Coffee and rice were also supported by special governmental institutions (Brazilian Institute for Coffee and Rio Grande do Sul Institute for Rice).

In the South, mainly in Paraná and Rio Grande do Sul, the introduction of mechanical technology in large-scale production of wheat, rice, and soya occurred simultaneously with a process of concentration in land distribution. In the north-east of Paraná, soya production, which is a capital-intensive system, was substituted for coffee, which was a labour-intensive and relatively small-scale activity. These changes magnified rural emigration during the 1970s and aggravated social problems in the region.

Agricultural growth and economic crisis in the 1980s

The Brazilian economy has grown rapidly since the end of the Second World War. During the 1980s, however, due to the accumulation of high external debts, a sudden increase in international interest rates, and an unfavourable evolution of the terms of trade, this trend was halted. The country's GDP increased by only 16.9% during the period, while in the previous nine years it had doubled (Table 2.3). Agricultural GDP presented a better performance; it increased by 28.2%, while the industrial sector was stagnant.

The adjustment programmes adopted by the federal government in the early 1980s affected credit provision and, consequently, the subsidized rural credit policy, the most important agricultural policy to date. Figure 2.2 shows that the amount of agricultural credit sharply increased in the 1970s, while in the following decade this trend was reversed. Up to 1979, the annual interest rate for operating rural credit was fixed at 15%. As the inflation rate during the period was higher than this level, the real interest rate was always negative. After 1979, the government introduced a mechanism to reduce progressively this implicit subsidy. The formula consisted of multiplying the inflation rate by a coefficient of 0.4 in order to obtain the rural credit interest rate. However, as inflation increased, the implicit subsidy continued to grow. In 1981, this method of indexation was substituted by a nominal interest rate of 45%. Again, due to high inflation rates, negative interest rates persisted up to 1983, although government rural credit provision was dramatically reduced in that year. In 1984 and 1985, a full mechanism of indexation was introduced and the subsidy was reduced to lower levels. In 1986, the Cruzado Plan set up rules to remove all mechanisms of monetary correction in the Brazilian economy; the rural credit interest rate was fixed at 10% nominal level. This change was conceived with the mistaken expectation of low inflation, which allowed for the return of the credit implicit subsidy in that year. In 1987, the full indexation mechanism was re-introduced and the real interest rates became positive.

As an alternative to credit constraints, the federal government reinforced the Minimum Price Guarantee Policy (PGPM). The basic instrument of the PGPM was the government announcement of the commodities guarantee prices at the beginning of the production cycle. Based on these prices, farmers could not only decide what to farm but also reduce uncertainty and economic risk. The 'minimum price' was the basic value that the government was prepared to pay for the commodity. Theoretically, after the harvest, farmers had three options: to sell their output in the market, to take a Federal Government Loan (EGF), or to sell their commodities through the Federal Government Purchase (AGF) schemes. Farmers, cooperatives, or intermediary agents could take up EGFs in order to support storage cost until market prices went up. At the end of the contract, if market prices were relatively low, they could switch from EGF to AGF, at the 'minimum price'.

Table 2.3
Indices of real production, Brazil: 1971-91, (1980=100)

Total GDP and main sectors

Years	GDP	*Agriculture*			*Industry*	*Services*
		Total	Crops	Livestock		
1971	48.7	69.4	71.9	66.8	45.9	45.3
1972	54.5	72.2	74.7	69.4	52.5	51.0
1973	62.1	72.2	75.7	67.4	61.4	58.9
1974	67.1	73.2	79.8	60.9	66.6	65.2
1975	70.6	78.0	82.9	70.2	69.9	68.4
1976	77.9	79.9	80.9	78.4	78.1	76.4
1977	81.7	89.6	90.8	87.6	80.5	80.2
1978	85.8	87.2	85.4	90.2	85.7	85.1
1979	91.6	91.3	90.8	92.1	91.5	91.7
1980	100.0	100.0	100.0	100.0	100.0	100.0
1981	95.6	108.0	109.6	105.5	91.2	97.8
1982	96.3	107.7	105.9	110.5	91.3	99.7
1983	93.0	107.2	104.0	112.2	85.9	98.9
1984	97.7	110.1	112.9	105.6	91.4	103.0
1985	105.7	120.6	127.8	109.5	99.5	109.7
1986	113.7	110.7	114.5	104.8	111.2	118.8
1987	117.8	127.5	132.4	120.0	112.3	122.7
1988	117.8	129.4	131.0	127.0	109.4	125.6
1989	121.7	133.1	135.2	129.9	112.6	130.4
1990	116.9	128.2	121.4	138.6	104.3	129.6
1991	117.9	130.9	123.7	142.1	103.5	132.2
1992	116.8	138.8	131.7	149.6	99.3	132.1

Sources: IBGE in Goldin and Resende (1993) and Central Bank of Brazil (1992)

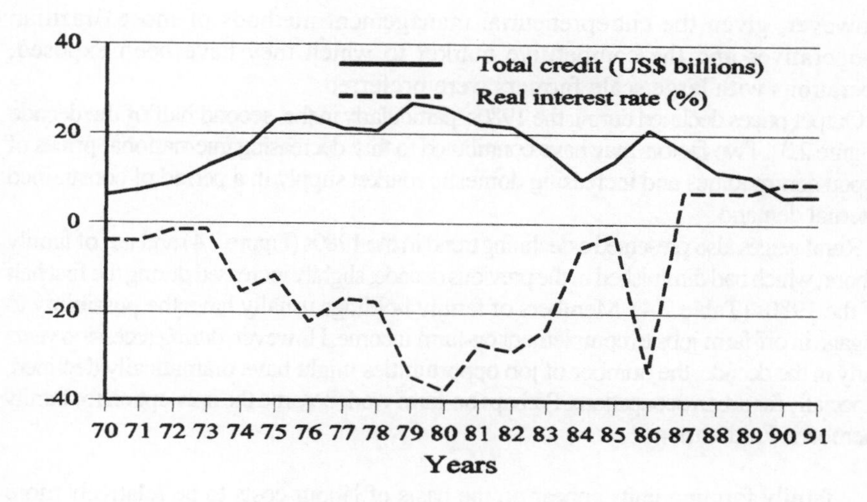

Figure 2.2 Rural credit loans and real interest rate, Brazil: 1970-91

In the early 1980s, the PGPM was not an important variable in farmers' decision making; generally, 'minimum prices' were fixed at low levels and government provision was relatively insignificant. Only after 1983, when credit constraints became severe, did the government fix prices and financial resources more realistically. This change increased the effectiveness of the policy. In fact, the Programme was one of the main factors explaining the relatively good performance of agricultural production during the second half of the decade. The scheme stimulated production of local food crops, such as rice, beans, and maize, which compensate for the distortions introduced by the rural credit policy during the 1970s, which favoured export activities. The 'minimum price' policy was specially advantageous to farmers who were located far from urban markets and faced high transportation costs. In this sense it was particularly important to stimulate the expansion of agricultural production into the new lands of the Centre-West savannas. Capitalized farmers, cooperatives, trade agents, and agro-industries received direct support from the Programme. However, most small farmers, mainly those who were not affiliated to cooperatives or associated with agro-industries, were not directly affected by the PGPM. The small farmers' operational scale was not compatible with the required commercial and financial structure. Moreover agro-industries and traders, most of them in a monopsonistic or oligopsonistic position, rarely transferred the government benefits to them. Theoretically, the PGPM could support small farmers through cooperatives.

However, given the entrepreneurial management methods of most Brazilian cooperatives and the competitive market to which they have been exposed, operations with large scale farmers were preferred.

Output prices declined during the 1980s, particularly in the second half of the decade (Figure 2.3). Two factors may have contributed to this: decreasing international prices of export commodities and increasing domestic market supply in a period of constrained internal demand.

Rural wages also presented a declining trend in the 1980s (Figure 2.4) and use of family labour, which had diminished in the previous decade, slightly increased during the first half of the 1980s (Table 2.4). Members of family holdings usually have the possibility to engage in off-farm jobs to complement on-farm income. However, during recession years early in the decade, the number of job opportunities might have dramatically declined, especially for urban occupations. Perhaps on-farm work became the only option for family members. Furthermore,

> family farming units appear on the basis of labour costs to be relatively more competitive during a recession; firstly, because the (inputed) wage for family labour may be expected to fall more sharply than for hired labour, whose wages - at least for some categories of workers - are related to those in the urban sector, and secondly, because whereas the cost of hired labour has to be fully covered, this variable cost does not constitute the same type of risk in family production, as it is the residual income (Goldin and Rezende, 1990, p.67).

The evolution of land prices could be linked to the phases of the economic cycle. As a whole, the price of land tended to decrease during the 1980s, although fluctuations have occurred because of changes not only in production related variables (rural credit and output prices), but also in external conditions (interest rate and inflation).[4]

Use of chemical fertilizers and crop protectors decreased in the early 1980s (Figure 2.5). In this period, changes in the rural credit policy could not be entirely taken as an explanatory variable for this decline; the implicit subsidy in the interest rate was still high and government credit provisions had not dramatically dropped (Figure 2.2). In fact, reduction in fertilizers and crop protectors consumption can be attributed to both the 1979 oil crisis, which triggered an increase in their prices (Figure 2.6), and a decrease in output prices (Figure 2.3). These conditions changed during the second half of the decade. Although implicit subsidies on rural credit were partially withdrawn, the PGPM became an important incentive policy and oil prices fell. Simultaneously, new land was brought into agriculture in the Brazilian savannas. Consequently, the consumption of fertilizers and crop protectors recovered to the levels of the early decade.

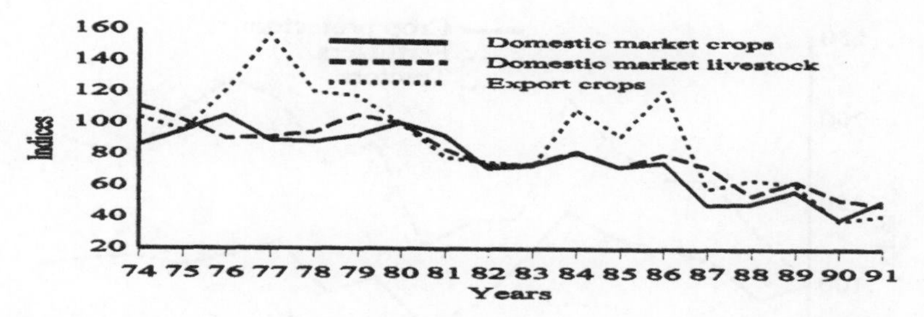

Figure 2.3 Indices of real prices received by farmers, Brazil: 1974-91 (1980=100)

Source: Goldin and Rezende (1993)

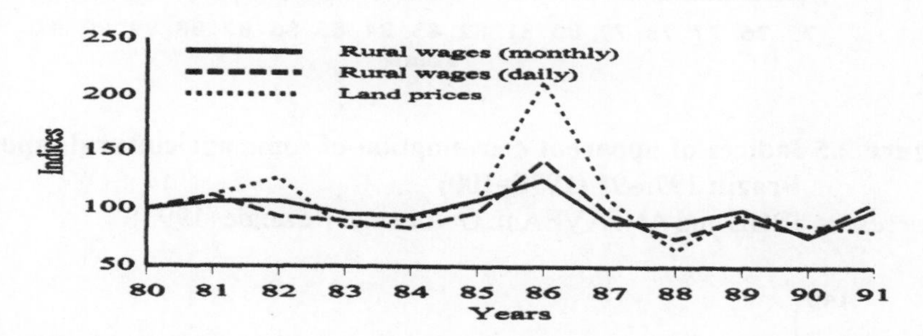

Figure 2.4 Indices of real rural wages and real land prices, Brazil: 1980-91 (1980=100)

Rural wages are deflated by IBGE's National Consumer Prices Index and land prices by the FGV's General Price Index-DI

Sources: Goldin and Rezende (1993) and Fundação Getúlio Vargas (1994 and 1995)

Table 2.4
Composition of the labour force, Brazil: 1970-85

Categories	1970	1975	1980	1985
Family	80	80	74	75
Hired Labour	15	16	23	21
Sharecropper*	5	4	3	3
Total	100	100	100	100

* Including other categories
Source: IBGE (1990)

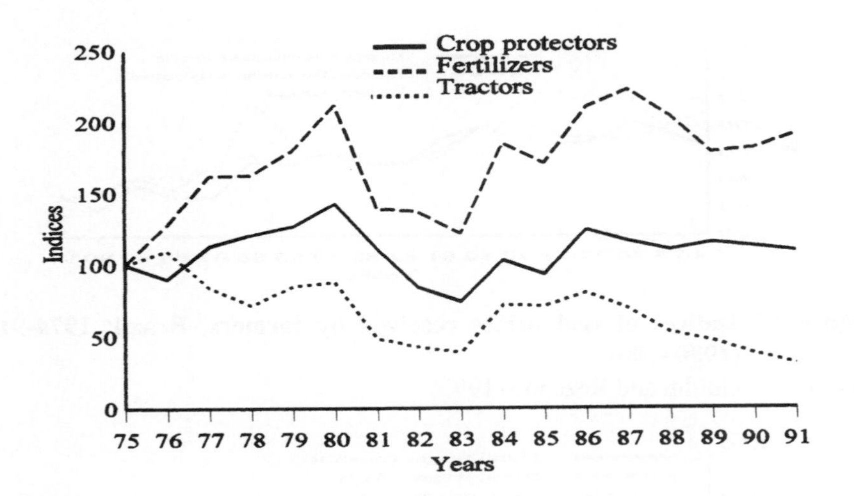

Figure 2.5 Indices of apparent consumption of some agricultural inputs, Brazil: 1975-91 (1975=100)

Sources: IBGE and ANFAVEA in Goldin and Rezende (1993)

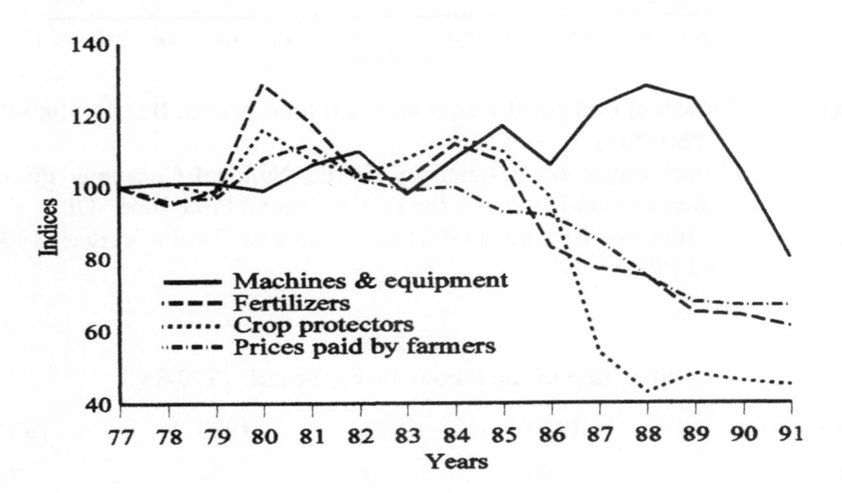

Figure 2.6 Indices of real prices of some selected inputs, and index of real prices paid by farmers, Brazil: 1977-91 (1977=100)

All indices were deflated by the IGP-DI

Sources: Goldin and Rezende (1993) and Fundação Getúlio Vargas (1994 and 1995)

Welfare in the Brazilian countryside

During the 1970s and 1980s, the distribution of rural credit in Brazil was highly selective. Because of recurrent deficits in the external balance, macroeconomic policy was adjusted to generate international trade surpluses. Internal demand was constrained and export and import-substitution activities (soya, coffee, sugar-cane, wheat) were stimulated. As most of these activities were concentrated in the South-East and South, these regions were favoured with a higher share in the national distribution of rural credit. Also, medium and large farms could obtain relatively large amounts of credit, while many small farmers were not covered by the government policies. During the 1970s, only 20% to 25% of Brazilian farmers were supported by official rural credit, and less than 5% were granted more than half of the total amount provided (Goldin and Rezende, 1990). This perverse scheme fostered the already latent process of income and land property concentration in the Brazilian countryside.

Hoffmann (1992) has calculated income distribution indicators in rural areas of Brazil. Table 2.5 shows that the mean income of the economically active population in the agricultural sector doubled between 1970 and 1980, although median income increased by only 58%. All indicators reveal a process of income concentration during the period. The indicator at the bottom of Table 2.5 shows that the incidence of poverty was reduced, although it was still high in 1980. If everyone's income had increased at the same rate as the population's mean income, the proportion of people below the poverty line would have been reduced. Moreover, inequity in land distribution was aggravated in this period (Table 2.6).

Table 2.7 shows that inequalities continued to be exacerbated during the 1980s. In 1986, as an effect of the Cruzado Plan which temporarily halted inflation, the proportion of the poor's income in the total income exceptionally increased. In the following years, however, the indicators show increasing inequality, which reached a peak in 1989, a year of extremely high inflation.

The values of the mean and the median income fluctuated during the 1980s. Between 1981 and 1983, when the economic recession was severe, these indicators showed a reduction of around 10%. The poverty indicator reached a local peak in 1983. After 1984, mean and median income increased, and the percentage of people below the poverty line decreased. In 1986, the number of job offers in urban areas increased due to the short-term effects of the Cruzado Plan. Consequently, supply of rural labour might have decreased, which allowed for the rural wages increase (Figure 2.4). However, after the Cruzado fiasco, the process of income concentration started again. The percentage of people below the poverty line reached 79.2% in 1990, which was worse than in the 1983 recession.

Apart from income and income distribution, other variables can be used to evaluate social welfare. Kageyama and Rehder (1993) have built an index to assess the standard of living in the Brazilian countryside in 1980 and 1990. They used data on twelve variables related to income and access to public services such as water supply, sewerage system, waste collection, education, electricity, social security, and health. All variables were measured in percentages,[5] which allowed for a weighted mean index. The weights were chosen to give priority to the social conditions that are relevant to the Brazilian context. For example, access to electric power is more important than public supply of water and waste collection; affiliation to the social security system is a highly meaningful component of the rural worker standard of living; and being above of poverty line is more important than the mean income level.

Table 2.5
Indicators of income distribution in agriculture, Brazil: 1970-80

Indicators		1970	1980
Mean income[a]		0.805	1.631
Median income[a]		0.580	0.916
Gini index		0.424	0.554
Proportion in	• 50% poorest[b]	23.6	17.4
the total income	• 10% richest[b]	35.0	48.8
of the:	• 5% richest[b]	25.7	38.8
Percentage of the poor[c]		81.4	59.5

a. As proportion of the real minimum wage of August 1980 (deflated by the Cost of Living Index - DIEESE).

b. The population was sorted by income in ascending order. Adding up the top observations up to 50% of the total number of observations and expressing it as a percentage of total income gives the proportion that is appropriated by the 50% poorest. Adding up the income of the bottom observations up to 5% (10%) of the total number of individuals and getting the percentage of the total income corresponding to this summation gives the proportion that is appropriated by the 5% (10%) richest.

c. Percentage of people whose income falls bellow the *poverty line* (value of the real minimum wage of August 1980).

Source: Demographic Censuses in Hoffmann (1992)

Table 2.6
Size distribution of rural properties, Brazil: 1960-85

Area (hectares)	Number of farmers					Area (%)				
	60	*70*	*75*	*80*	*85*	*60*	*70*	*75*	*80*	*85*
Less than 10	44.7	51.2	52.1	50.4	52.9	3.4	3.1	2.8	2.5	2.7
10 to 100	44.6	39.3	38.0	39.1	37.1	19.0	20.4	18.6	17.6	18.5
100 to 1,000	9.4	8.4	8.9	9.4	8.9	34.4	37.0	35.8	34.7	35.1
1,000+	1.3	1.1	1.0	0.9	1.1	44.2	39.5	42.8	45.0	43.7

Sources: Agricultural Censuses in Helmar (1994) and Shiki (1991)

Table 2.7
Indicators of income distribution in agriculture, Brazil:[a] 1981-90

Years	Mean income[b]	Median income[b]	Gini	50% poorest[c]	10% richest[c]	5% richest[c]	% of the poor[d]
1981	1.02	0.60	0.659	7.3	49.6	36.1	0.70
1982	0.98	0.52	0.657	7.4	49.3	36.7	0.73
1983	0.93	0.54	0.674	7.1	52.3	38.9	0.74
1984	0.96	0.54	0.671	7.7	52.3	38.6	0.74
1985	1.03	0.55	0.682	6.9	53.5	39.7	0.73
1986	1.48	0.79	0.661	8.4	51.0	37.6	0.61
1987	1.04	0.54	0.682	6.9	52.9	39.4	0.71
1988	0.98	0.52	0.677	7.2	53.1	39.3	0.74
1989	1.15	0.55	0.697	6.4	54.9	41.3	0.70
1990	0.83	0.43	0.682	7.0	52.6	39.1	0.79

a. Excluding North Region.
b. As proportion of the real minimum wage of August 1980 (deflated by the INPC).
c. See Table 2.5, note b.
d. See Table 2.5, note c.

Source: PNAD in Hoffmann (1992)

Table 2.8
Indices of rural welfare, Brazil and 20 States: 1981 and 1990

States	1981	1990	Rel. prog.*
North-East			
Maranhão	23.7	29.0	6.9
Piauí	15.7	18.0	2.7
Ceará	22.7	24.2	1.9
Rio Grande do Norte	26.0	30.8	6.5
Paraíba	20.3	23.4	3.9
Pernambuco	31.6	36.5	7.2
Alagoas	32.0	31.6	-0.6
Sergipe	28.8	34.1	7.4
Bahia	31.4	31.3	-0.2
South-East			
Minas Gerais	39.4	46.6	11.9
Espírito Santo	40.5	45.2	7.9
Rio de Janeiro	53.8	58.6	10.4
São Paulo	60.1	71.7	29.1
South			
Paraná	45.1	54.2	16.6
Santa Catarina	54.6	62.3	17.0
Rio Grande do Sul	51.9	61.2	19.3
Centre-West			
Mato Grosso do Sul	45.9	57.6	21.6
Mato Grosso	41.5	52.2	18.3
Goiás	39.1	47.2	13.3
Distrito Federal	53.8	63.6	21.2
Brazil	38.8	43.5	7.7

* Relative progress = [(1990 value − 1981 value)/(100 − 1981 value)] x 100

Source: Kageyama and Rehder (1993)

Table 2.9
Rural, urban, and total population (millions), Brazil: 1970-90

Areas	1970		1980		1990	
	Population	*%*	*Population*	*%*	*Population*	*%*
Rural	41.1	44	38.6	32	36.0	25
Urban	52.0	56	80.0	68	107.7	75
Total	93.1	100	119.0	100	143.7	100

Source: IBGE (1990) and IBGE (1991)

The results are presented in Table 2.8. Considering that the indices can lie between 0 and 100, the outcomes are very low, which reveals the low standard of living of the rural population. The evolution during the decade was slightly positive for the country as a whole, though the differences among States reveal extreme heterogeneity. The rates of relative progress for the most developed States of the South and South-East were higher than the ones of the North-East States.[6] The greater the welfare in the early period, the greater the rate of relative progress. Kageyama and Rehder's indices confirm that Brazil's economic growth tended to favour those who already had privileged positions.

The drawback of this analysis, as the authors recognized, is that the data only refer to the rural population. People who migrated to urban areas during the period were not considered. Most have moved due to the same factors that allowed for improvements in the welfare of those who remained in the countryside. The rural-urban flow was estimated at 16 million during the 1970s and 10 million in the following decade (Mueller and Martine, 1994). The rural population decreased in this period (Table 2.9). The reduction in the flow during the 1980s is explained by two main factors: a decrease in the rate of births, which had been declining since the 1960s; and the economic instability. Mueller and Martine have suggested that the 1980s economic crisis had two effects on the migration process. First, as the rate of unemployment increased, potential migrants, generally relatively well informed people, gave up their intention to move. It became easier to survive in the area of origin, where network contacts and informal schemes of mutual aid were better known. Second, lack of financial resources constrained both diffusion of large-scale technologies and land speculation, which had contributed to the higher migration flows of the previous decade.

Effects on environment and human health

Given the great diversity of natural environments and the regional differences in terms of technological intensification, Brazilian agriculture presents a complex range of environmental concerns. For instance, the problems in the South and South-East Regions are similar to those of the developed countries, such as water pollution by pesticides, erosion, soil compaction, decreasing fertility, and chronic plant disease problems. In the North, on the other hand, the problems are related to the chaotic expansion into a fragile environment (Flores et al., 1991). An overview of the matter is provided by EMBRAPA (1993). In this document, the national territory is divided into six ecological regions (the Amazon Forest, the Central Savannas, the North-Eastern Semi-Arid, the Atlantic Forest, the Southern Forests and Grassland, and the Mato Grosso's *Pantanal*), for which the main environmental problems are summarized in Table 2.10.

Table 2.10
Environmental problems in the Brazilian ecological regions

Amazon Forest Farming systems in this region can be characterized as low-input and unsustainable. Farmers' level of education and technological information is low; and insects and diseases exerted strong biological pressures on production. The techniques and crops brought by immigrants from the South, South-East, and North-East were not always appropriate for local conditions, therefore contributing to worse environmental problems. In general, the colonization process was disordered and predatory: deforestation, indiscriminate burning, soil erosion, genetic resources losses, landholders conflicts, and lack of infrastructure and basic services. Migratory or itinerant agriculture is the most common farming method in the equatorial Amazon area. The system is characterized by the cutting and burning of trees in small isolated areas, where farming is sometimes restricted to a maximum period of two years, time in which the ashes are still working as fertilizer and soil corrective. This is followed by a fallow period of around eight years, which has been reduced in areas of greater population density. Substitution of riverside forests for pastures along *varzeas* (wetlands) of great rivers has brought serious environmental problems. Deforestation in areas where iron industries, large hydroelectric dams, and mineral prospecting activities have been established, has caused great concern. Most of these cleared areas have been devoted to cattle raising. Because of soil low fertility and inadequate management, these pastures remain productive for only five to eight years; afterwards, they are abandoned due to high level of degradation.

Central Savannas In spite of the predominance of chemically poor soils, the gently rolling topography of the region, along with other excellent physical characteristics and the availability of water, makes a large part of their 204 million hectares similar to the arable area of the USA, and highly favourable for agriculture. In areas where large agricultural projects were set up, inadequate use of technologies (mechanization, irrigation, chemical inputs) and soil management have caused compaction of soils, erosion and loss of fertility. The process of uncontrolled agricultural expansion in the region must necessarily be contained and land use should be planned according to delineation of agri-ecological zones.

North-Eastern Semi-Arid During severe dry seasons, many farmers in the Semi-Arid region clear the remaining vegetation for survival. They sell firewood and charcoal, or migrate to more favourable regions. In the irrigated areas, the main environmental problem is soil salination, which is a consequence of inadequate water management. During the rainy season, showers cause heavy soil erosion, especially due to lack of protective green cover. Alternations between dry periods and catastrophic floods are common. Both surface and underground waters are strongly affected by salts, which makes irrigation difficult.

Atlantic Forest There are many ecological problems along the Atlantic Coast. In the past, the Atlantic Forest formed an uninterrupted strip of rainforest from the State of Rio Grande do Sul to the State of Rio Grande do Norte. It is one of the most threatened ecosystems in the world. Only 7% of its original area still remains unexploited; most are in Official Conservation Reserves or in steep hills. Even so, charcoal dealers and lumber merchants assaults are a constant threat. The soils are in general of medium fertility, but intensive agriculture has been limited by the rough ground. Most of the lower areas, which have more fertile soils, have already been cleared. In the North-East, where the Atlantic Forest has almost been totally cleared, sugar-cane monoculture has caused not only serious environmental impacts, but also negative social consequences, as they are located in remarkably suitable food production lands.

Southern Forest and Grassland In general, the soils in this region are naturally fertile, which, with an amenable climate, made the last century's rapid colonization possible. Inadequate systems of soil preparation have characterised the agriculture of the region. Intensive mechanization and expansion of cereals cultivation have contributed to both loss of organic matter and pulverization/compaction of soils, which generate water and wind erosion. Besides, lack of green cover has interfered in the natural accumulation of underground water. Slurry from intensive pig production has been discharged into rivers and lakes, resulting in serious problems of superficial water pollution. The *Campanha* grassland area is extremely suitable for beef cattle rasing, although management is inadequate. There is no concern with recovery and maintenance of the vegetation, and the grassland is used under continuous and extensive grazing. The problem is aggravated by excessive cattle trampling.

Mato Grosso's Pantanal The *Pantanal* is a sedimentary upland within the Basin of the Paraguay River. The complex hydrological system and its dynamic have decisive influence on the region's biodiversity and production activities. Most soils of the Pantanal are formed by sedimentation of materials eroded in the adjacent plateau. Most have sandy texture and low natural fertility. The fauna is rich and diversified, although in recent years illegal hunting and fishing have threatened the system. Forests have been cleared to allow for subsistence cropping, cultivated pasture, irrigation projects, dikes, dams and roads. The main economic activity is beef cattle raising. In the plateau that surrounds the Pantanal, large scale and intensive cereals farming, which is often practised on sandy soils, has been provoking ecological imbalances and carrying away sediments that have affected the hydrological system in the plain areas.

Source: EMBRAPA (1993)

Environmental pollution due to crop protectors and fertilizers in Brazil has been studied by several governmental and non-governmental organizations. However, lack of systematic and quantitative data, and of evaluations at a national level, have limited a comprehensive analysis of the problem. Most studies are site specific and have focused on the effects of particular hazardous chemicals on farm workers, food, and water sources. Some of these studies were reviewed by Margulis (1988), who also interviewed 95 experts in Brazil during the period March-July 1986. Surveys among pesticide operators in the States of Paraná and Rio de Janeiro showed that misuse has caused human poisonings and environmental pollution. A high percentage of operators did not follow agronomic prescriptions, lacked adequate protective clothing, smoked during the applications, stored the products under unsafe conditions, and washed their equipment in rivers, tanks, wells or lakes. Although the statistics were site specific, all experts admitted that these circumstances were very representative of the situation of pesticides use in Brazil; a case of low level of training and education of workers and/or inadequate working conditions.

The issue of pesticide residues in food began to receive closer attention in the early 1970s, when some export products were analysed by importing countries and embargoed because they were considered contaminated. As a result, new laboratories were created and the old ones were modernized. In addition, legal standards for pesticide residues in agricultural products were established. The studies reviewed by Margulis (1988) suggested that the level of residues in food cannot be considered high in Brazil, a view shared by the experts he interviewed in the country. The 1985 ban on the use of organochlorines may have further decreased food contamination. Studies on water pollution also revealed low levels of pesticides, although none of them was carried out in the State of São Paulo, the most important State in terms of pesticide use.

A more comprehensive list of accidents, medical reports, case studies of specific geographic areas, and export embargoes on contaminated foodstuff is provided by Henão et al. (1991). The document was prepared for the Ministry of Health and comprises a preliminary analysis of the use of agrochemicals and their impacts on human health and environment in Brazil. Since the 1960s, pesticide residues have been found in several agricultural products: fruits, vegetables, potatoes, wheat, milk, beef, and canned beef. BHC residues were also detected in fish, shrimps, and oysters in São Paulo's coast. There are several cases of poisoning of farm workers by pesticides (on cotton in Paraná, cocoa in Bahia, and sugar-cane in Rio de Janeiro). Cases of herbicide spraying destroying neighbouring plantations due to wind action were reported. Furthermore, hazardous chemical residues were detected by soil analysis in São Paulo.

Public concern about those kind of problems has increased since the early

1970s. In 1971, the first Brazilian ecological association, the AGAPAN (*Gaúcha* Association for Natural Environment Protection), was created and has denounced the environmental damage caused by indiscriminate use of agricultural pesticides (Paulino, 1993). After 1974, because of the military government's relaxation of controls, other pro-ecology and non-governmental organizations were created in several cities of the South and South-East Regions.

In response to this public concern, governmental agencies were established in some States to deal with environmental problems, and some politicians were elected in the 1982 elections by including environmental issues in their platform. The country's political reform and the debates that led to the 1988 new Constitution established new grounds for the discussion of environmental issues. Urban and rural ecologists worked together in favour of the institution of agrochemical regulatory laws in the States, which culminated with the 1990 national legislation on pesticides (Table 2.11).

Since 1990 the country has had more restrictive legislation, although existing technical and administrative capabilities have been insufficient to enforce it (Silveira, 1994). Laboratory facilities are inadequate to develop eco-toxity tests and to evaluate products' chemical qualities. Control and legislative functions follow a bureaucratic process through three governmental agencies: the Ministry of Agriculture, which analyses the biological value of pesticides; the Ministry of Health, which analyses the pesticide content in foodstuffs and its effects on health; and IBAMA (Brazilian Institute for the Environment), which deals with the eco-toxicological effects. The procedure for registering a product is expensive and can take three to four years to complete. Given these problems, lobbies and interest groups have exerted strong pressure to relax restrictions and undermine effectiveness of the law.[7]

Governmental and non-governmental efforts towards sustainable agricultural technologies

At governmental level, initiatives towards research and diffusion of agricultural sustainable technologies in Brazil have been carried out by EMBRAPA (Brazilian Agricultural Research Corporation). This State research Corporation was created in 1972 to support the Green Revolution technological changes. It has been considered successful in terms of technology development, although increases in production and productivity have not always contributed to solving social problems in the country's rural areas. Since 1985, when the civil government succeeded the military regime, social and environmental targets were incorporated in the EMBRAPA's strategy of development (Almeida, 1989). In terms of research, more emphasis has been given to basic foodstuffs production, conservation, environmental quality, fossil-fuel saving technologies, and development of site-specific technologies.[8]

> In its historical context, Brazilian agricultural research has traditionally been oriented towards the production of technologies capable of producing high yields, not always with a concern for environmental protection, which has now become part of the new paradigm for the Corporation (EMBRAPA, 1993, p.24).

Table 2.11
Chronology of pesticide regulation in Brazil

1934 The first Brazilian law on pesticides was created. At that time, restrictions on the use of organochlorines was not included in the law.

1972 The R.A. (agronomic prescription) was informally established as a prerequisite for pesticide purchase in Santa Rosa, a municipality of the State of Rio Grande do Sul.

1978 The Rio Grande do Sul branchs of the Bank of Brazil established the R.A. as a prerequisite for pesticide purchases.

1980 The Bank of Brazil extended the R.A. prerequisite to other States.

1981 The R.A. prerequisite is instituted by the Ministry of Agriculture for all pesticide purchases in the country.

1982 The government of the State of Rio Grande do Sul banned organochlorines products and created legislation on pesticide trade. The latter was subsequently followed by other States.

1985 Organochlorines were banned in all Brazilian territory.

1988 The new Brazilian Constitution was approved by the Congress and institutes laws on environmental conservation. 'Agrression on the environment' can be legally punished and the public sector has the duty to regulate production, trading, and the use of techniques, methods, and materials that bring risk for life, quality of life, and the environment.

1990 A new and comprehensive law on research, experimentation, production, packing, labeling, transportation, stocking, trading, advertisement, use, importation, exportation, residues and package disposal, registration, classification, control, and inspection of pesticides and their components was sanctioned by the President.

Source: Paulino (1993)

EMBRAPA and other governmental institutions have exerted great effort towards biological plant protectors and development of IPM (Integrated Pest Management) techniques. An informal partnership has involved the Corporation and cooperatives, especially in programmes for soybeans and wheat (Silveira, 1994).

Although governmental programmes have shown concern with environmental problems, their success in attaining social goals is dubious. They have been carried out under the same institutional arrangement that in the recent past allowed the negative social effects of the Green Revolution. If IPM programmes fail to extend their benefits to the small farmers who were marginalised by government policies in the last decades, they will not help to reverse the process of income concentration in rural areas.

A broader concept of sustainable development has been adopted by some non-governmental organizations, which have been diffusing LEISA (low-external-input and sustainable agriculture) practices through Participatory Technology Development (chapter 3).[9] They have been founded by environmentalists, unions, cooperatives, small farmers' associations, churches, and political activists. Some have clear environmental objectives, while others are politically oriented. There is a predominant concern with the survival of the smallholding and a strong antagonism to the Green Revolution. Table 2.12 shows some centres and non-governmental organizations that deal with the development and extension of sustainable agricultural technologies in Brazil.

The 1980s democratization process allowed most of these organizations to channel rural social demands. The diffusion of 'alternative technologies' has been used as a strategy for action (von der Weid and Almeida, 1988). Some politically oriented organizations recognised the need to associate the agrarian reform campaign with solutions for farmers' immediate production and commercialization problems. 'Alternative technologies' were seen as a tool to strengthen the smallholders' economic power in their struggle to remain on land (von der Weid and Almeida, 1988). For politically oriented organizations, solutions to immediate social problems are sometimes more important than following the strict technical prescriptions of some ecologically oriented groups.[10]

Conclusion

This chapter has shown that the Green Revolution in Brazil has failed to achieve important objectives of sustainable development (chapter 4). Depletion of the natural environment and inequalities among farmers and regions have been exacerbated. On the one hand, both the introduction of inadequate technological packages and the expansion of agriculture into valuable natural ecosystems have

Table 2.12
Some organizations and centres that deal with alternative technologies in Brazil

CAPA Small Farmers' Advice Centre, State of Rio Grande do Sul. The Centre was founded in 1978 by the Lutheran Church.

ASSESSOAR Studies, Orientation and Assistance Association, Francisco Beltrão, State of Paraná.

MOC Feira de Santana Community Organization Movement, State of Bahia.

PATAC Community Adapted Technologies Application Programme, Campina Grande. State of Paraíba.

AS-PTA Alternative Agriculture Project Consultants, Rio de Janeiro. AS-PTA and other 15 organizations make up the Alternative Technology Project Network (Rede PTA).

Estância Demétria (Demeter Farm) Botucatu, State of São Paulo. The farm is supported by the Biodynamic Institute for Rural Development, which provides certification for biodynamic and organic agriculture.

Sao João Baptista Vianei Institute Vianei Education Project, Lages, State of Santa Cantarina.

Augusto Ruschi Ecological School Horticultural experimentation station in Cachoeiro de Itapemirim, State of Espírito Santo.

CIER Rural Education Integrated Centre, Boa Esperança, State of Espírito Santo.

TAPS Brazilian Association of Alternative Technology for Health Promotion, State of São Paulo.

CETAP Centre for People's Alternative Technologies. CETAP was founded in 1986 by the Landless Movement, rural unions, and the Farmers Affected by the Reservoirs Movement .

CTA Oricuri Alternative Technologies Centre, State of Pernambuco.

CTC Cotrijuí Training Centre, Ijuí, State of Rio Grande do Sul. The Centre is supported by the Cotrijuí (Ijuí Wheat Growers Cooperative).

CAMP Multiprofessional Advising Centre, Porto Alegre, State of Rio Grande do Sul.

MEPS Educational and Promotional Movement of Espírito Santo, State of Espírito Santo.

Sources: Henão (1991), Instituto Biodinâmico (1995), AS-PTA(1994), and Programa de Cooperação em Agroecologia (1992)

caused serious environmental problems (biodiversity reduction, soil depletion, water pollution, and salination). Consequently, the possibility of achieving intergenerational equity in the distribution of natural resources was lost. On the other hand, income distribution and welfare indicators have shown that the economic growth has favoured those who were already in privileged positions.Therefore, inequalities among people in the same generation (intragenerational distribution of wealth) have widened. Poverty in the Brazilian agricultural sector has been fostered by two factors: the diffusion of high-external and unsustainable technologies over a highly concentrated agrarian structure and a social structure that favoured the elites; and the persistence of traditional, destructive farming systems. Both kind of practices deplete natural resources and reduces potential sources of income for present and future generations. Although the government and non-governmental organizations have taken positive steps to remedy these problems (including legal restrictions on crop protectors use, fine tuning of conventional technologies, and diffusion of LEISA practices), these are still insufficient to provide solutions.

Notes

1 For export activities, such as oranges, sugar-cane, tobacco and coffee, physical yields increased at annual rates of 3% to 5% between 1971 and 1980, while for internal market products, such as maize, rice and beans, these rates were negative (Goldin and Rezende, 1993).
2 A broader view has been given by Kageyama (1986).
3 Many large-scale projects were officially approved under the justification that 25,000 ha was the minimum size required both to make extensive cattle grazing farming a profitable activity in the region and to preserve the ecological balance (50% of the land of approved projects should be kept as natural forest reserve).
4 Land in Brazil is simultaneously a production and a liquid asset. Therefore, its price is affected by economic variables that are not directly related to agricultural production (Romeiro and Reydon, 1994).
5 The only exception was mean income, for which a conversion method was devised.
6 The change in the welfare indices between 1981 and 1990 was calculated according to the idea of 'relative progress'. Given that there is a maximum limit for the Indices of Rural Welfare, in the States where the initial value is already high, the changes would be necessarily small. This problem restricts comparisons between States with large initial differences. To

remedy this Kageyama and Rehder used the idea of variation in relation to the maximum possible change, which is expressed by: (value of the 1990 index - value of the 1981 index)/(100 - value of the 1981 index) . The numerator expresses the actual growth achieved in the period, while the denominator expresses the maximum growth that could be achieved.

7 Pesticide industries have launched new and supposedly less aggressive products in the market, along with campaigns to rationalize pesticide use (Paulino, 1993). Most are multinational companies, which have experienced developed countries' restrictions.

8 In 1993, EMBRAPA had 39 research units (EMBRAPA, 1993), of which four were directly dealing with IPM and biological control: CNPSo (National Soybean Research Centre), CNPT (National Wheat Research Centre), CENARGEN (National Genetic Resource and Biotechnology Research Centre), and CNPMA (National Research Centre for Monitoring and Assessment of Environmental Impact) (Silveira, 1994).

9 An attempt to account to their wide range of technical and philosophical orientation is given by Almeida (1989).

10 Non-governmental organizations have diffused different methods of organic production in the country; some have certification schemes, others do not. This situation has led the government to call representatives of several non-governmental organizations to jointly discuss a new law to regulate production, processing and certification schemes for organic products in Brazil.

3 Economic, social, and environmental aspects of the Green Revolution in the State of Espírito Santo

Introduction

The Green Revolution in Brazil was examined in chapter 2. It was evident that although the changes have followed the same direction in all Brazilian States, regional characteristics affected speed, intensity, and consequences. The objective of this chapter is to examine the specific elements of the Green Revolution in the State of Espírito Santo, and to provide a background for the empirical analysis of chapter 8. The historical development of Espírito Santo's agrarian sector, the economic and social changes that followed the introduction of the mechanical and chemical-based technologies, and the effects on environment and human health are examined here.

Historical antecedents[1]

The history of agriculture in the State of Espírito Santo is connected with the history of Brazilian coffee, which became important in the world market in the middle of the nineteenth century when French production in Haiti collapsed. Slave plantations were initially established in the mountain areas of Rio de Janeiro and expanded extensively until they reached the neighbour Provinces of Minas Gerais and Espírito Santo.

Before the introduction of coffee, sugar-cane had been the main agricultural product in Espirito Santo. Slave plantations were established in the plains areas that lie between the *Serra do Mar*, a mountain chain, and the coast. By the turn of the eighteenth century, however, prices had fallen and production declined. With the expansion of Rio de Janeiro's coffee production, farmers of the State of Espiríto Santo became interested in this product and established plantations to substitute for sugar-cane. The dense rainforest and the presence of river rapids made it difficult to establish farms in the hinterland, where altitude and weather conditions would be more appropriate for coffee production. Only after 1870, when coffee prices increased and Rio de Janeiro's railway system reached Espírito Santo's border, were plantations established in the mountains.[2]

By the end of the nineteenth century, only 15% of the territory of Espírito Santo had been occupied by agriculture, mainly with coffee and sugar-cane in the South; the remaining area was unexplored Atlantic Forest (Almada, 1984). To occupy the territory economically and provide labour supply to the former slavery plantations, the government of Espírito Santo encouraged European immigration.[3] New settlements, which would be characterized by the predominance of subsistence small-holdings, were established in the mountainous region on the West of Vitória, the capital of the State. The proximity of the port of Vitória and increasing prices led local dealers to encourage coffee farming among settlers. The production of this commodity would provide them with a monetary surplus to buy off-farm goods. In this mercantile structure, dealers would occupy both monopolistic and monopsonistic positions.

Between 1897 and 1910, and 1913 and 1918, coffee prices sharply decreased. Although this reduced settlers' monetary income, the subsistence system did not collapse. As the production of basic foodstuffs was independently undertaken, survival was guaranteed. Moreover, coffee fields were not abandoned, as coffee, a permanent crop, yielded annual harvests without heavy, monetary cost. The same cannot be said about the large slavery plantations. After the 1888 abolition of slavery, farmers started to employ European immigrants, mostly Italians and Germans, as sharecroppers. Falling coffee prices led to further economic distress and many landowners sold off plots of land. Some of these were bought by sharecroppers, who also continued supplying temporary labour to the remaining plantations.

The sugar-cane plantations were also in economic difficulties due to the abolition of slavery and declining output prices. Some farmers tried to switch to coffee, but, in areas close to the coast, climatic conditions were not favourable. The crisis helped the fragmentation of large farms in the region.

Espírito Santo became a State in which small holdings predominated. According to the 1920 Brazilian General Census, 89% of the number of farms (comprising 52% of the total area of farms) had less than 100 hectares. Only 0.3% of farms, acounting for 10% of the total area, had more than 1,000 hectares. Most of the territory in the State, however, was still uncultivated (71%), and, within the farms themselves, the percentage of farmed land was only 18%, distributed across coffee (68%), sugar-cane (5%), and subsistence crops (27%), such as maize, beans, cassava, and rice.

Up to the 1920s, the region to the North of the River Doce, which comprised

55% of the territory of Espírito Santo had little involvement in agricultural activities. There were a few cocoa plantations on the margins of the River, and, in the extreme North, the Village of São Mateus and the farms around it formed a cluster of cassava and cassava-flour production. As the population continued to increase and the availability of land in the South and Central areas of the State was reduced, new settlements were created in the North. The coffee-subsistence smallholding farming system expanded up to the second half of the 1950s. Urban markets increased and transport facilities were improved, creating conditions for other products, such as cattle, hardwood, beans, rice, and maize, to be brought into commercial production, although coffee remained the main commodity.

After 1955, the international price of coffee fell sharply due to overproduction. Similar crises had happened before, but the federal government had always bought up excess production to sustain domestic prices and avoid economic depression. In the 1960s, however, the government decided to introduce a new scheme to reduce production. Between 1962 and 1967, through two Eradication Programmes, farmers were offered a compensation payment to destroy coffee plants. As a result more than 1.3 billion plants (1.6 million hectares) were eradicated in Brazil during this period. In Espírito Santo, 0.4 billion plants (0.3 million hectares) were destroyed (Guarnieri, 1979). The policy was devastating for the local economy, which was then largely based on coffee production. Around 20% of the rural work force was made redundant. The government provided credit for diversification to alleviate social problems and increase food supply. However, profitability of local market crops was low and soils were depleted. Most land was re-allocated to cattle grazing, a labour-saving activity, which exacerbated the inequities in land distribution and fostered rural-urban migration.

The Green Revolution[4]

In the most developed regions of Brazil, the Green Revolution had started by the end of the 1960s (chapter 2). In Espírito Santo, however, the Eradication Programmes and the low prices of coffee delayed the mass introduction of the technological packages to the second half of the 1970s, when agricultural production increased sharply.

Subsidized rural credit was the most important policy fostering the Green Revolution. Between 1970 and 1980, the total amount of rural credit in the State increased by five fold (Table 3.1). Other agricultural programmes were also set up and contributed to increased production and promoted technical change. Table 3.2 and Table 3.3 show the rate of growth of main agricultural activities and the evolution of land use, respectively.

Indicators of technological intensification, and the evolution of the agrarian structure are presented in Table 3.4. Diffusion of Green Revolution technologies accelerated between 1975 and 1980 (indicators 1, 2, 3, and 4), and farms became more dependant on the use of external (purchased) inputs (indicator 5). In this process, the participation of family labour in total labour decreased, while the percentage of hired, mainly seasonal, labour increased (indicators 6, 7, and 8).

Table 3.1
Evolution of rural credit in Espírito Santo: 1969-93 (1980=100)

Years	Total	Crops	Livestock	Coffee	Coffee/ total (%)
1969	15.7	12.4	25.4	2.8	6
1970	19.5	12.7	39.8	1.9	3
1971	24.8	13.6	58.2	3.9	5
1972	28.2	17.3	60.6	6.6	8
1973	35.8	21.1	79.5	8.9	8
1974	46.3	29.5	96.6	12.4	9
1975	76.8	39.6	188.0	25.4	11
1976	80.0	44.7	185.5	33.4	14
1977	67.2	53.7	107.4	41.5	21
1978	70.1	51.6	125.5	46.8	22
1979	89.5	70.2	147.1	80.0	30
1980	100.0	100.0	100.0	100.0	33
1981	71.3	75.3	59.6	71.4	33
1982	90.7	100.6	61.3	114.0	42
1983	91.3	101.4	61.1	119.7	44
1984	50.4	59.2	24.2	65.8	44
1985	67.3	80.7	27.1	94.2	47
1986	88.2	101.5	48.4	104.7	40
1987	69.4	83.6	26.7	111.2	53
1988	22.3	28.5	3.6	33.7	50
1989	17.4	22.3	2.8	24.0	46
1990	12.4	15.1	4.5	16.0	43
1991	9.9	11.9	3.9	4.7	16
1992	7.5	8.1	6.0	3.0	13
1993	3.6	3.8	2.8	0.8	7

Sources: CONCRED/MA and Banco Central in Souza Filho (1990); and Banco Central

Table 3.2
Geometric rates of growth of crops and livestock production, Espírito Santo: 1960-92

Products	60-65	65-70	70-75	75-80	80-85	85-90	90-92
Crops							
Pineapples	31.2	35.8	15.3	-16.5	15.8	-	-
Garlic	-	3.3	-15.7	34.4	29.1	16.8	-7.4
Rice	11.1	4.2	-3.4	-2.0	11.5	-2.2	-5.6
Bananas	10.2	6.1	-1.7	-11.2	1.5	-2.5	19.0
Potatoes	-1.2	16.5	0.6	-21.3	20.7	18.9	-17.5
Cocoa	-10.1	12.4	4.0	11.2	1.6	-	-
Coffee	-3.5	-9.7	-3.4	23.5	12.4	-3.4	8.4
Sugar-cane	5.0	-0.7	-3.9	4.4	31.7	na	na
Beans	-0.6	5.1	1.5	1.5	-4.3	11.6	-5.5
Oranges	5.4	2.4	8.1	-21.4	5.7	-3.9	13.3
Papaya	-	-	-	15.2	83.1	-	-
Cassava	5.0	11.2	-7.0	-6.8	5.7	-10.4	-5.4
Maize	2.2	11.2	-5.1	2.3	0.3	-3.0	19.5
Black-pepper	-	1.8	20.3	20.4	39.8	-	-
Tomato	-	37.0	23.5	7.0	8.7	8.2	-3.6
Livestock							
Hens	8.9	2.8	-7.9	5.0	-10.3	-2.1	-0.3
Cattle	7.4	7.3	4.8	-2.8	-1.9	-0.4	4.8
Pigs	3.7	1.4	-14.6	-5.2	-1.6	-0.4	-0.7
Milk	14.1	3.1	10.1	2.8	1.1	-	-

Sources: IBGE and CEPA/ES in Souza Filho; and IBGE

Table 3.3
Evolution of land use in Espírito Santo's agriculture: 1960-85

% of the total area in use

	1960	*1970*	*1975*	*1980*	*1985*
Permanent Crops	18	9	10	16	20
Temporary Crops	12	12	10	8	10
Pasture Land	34	57	64	59	54
Natural Forests	35	20	13	13	11
Reafforestation	1	1	3	4	5
Total Area in Use	100	100	100	100	100
Total Area in Use (million ha)	2.46	3.20	3.32	3.37	3.51

Sources: IBGE in Souza Filho (1990); and IBGE (1985)

Table 3.4
Some statistical indicators of the Green Revolution in Espírito Santo: 1960-85

	1960	1970	1975	1980	1985
1 Land/Tractor (ha)	4,850	2,833	1,713	631	368
2 Tractors/1,000 rural workers	-	3	6	14	21
3 Farms using chemical fertilisers (%)	-	1.9	17.1	50.8	57.7
4 Farms using chemical pesticides (%)	-	-	41.8	57.3	47.8
5 Monetary expenditure/ha[a]	-	100	103	201	325
6 Family labour/Total labour (%)	-	60.3	58.6	47.3	43.1
7 Permanent hired labour/Total (%)	-	5.9	10.3	14.1	12.1
8 Seasonal hired labour/Total (%)	-	13.7	13.3	19.1	21.1
9 Mean farm size (ha)	52.7	53.2	63.4	63.9	56.3
10 Land distribution (Gini)	0.550	0.604	0.628	0.654	0.673
11 The smallest 50% (share of the area)[b]	15.8	12.7	11.8	10.7	9.6
12 The largest 5% (share of the area)[b]	33.1	37.2	40.5	43.3	44.8

a Index 1970 = 100.

b Farms were sorted by area in ascending order. Adding up the area of the top observations up to 50% of the total number of farms and expressing it as a percentage of total area gives (11). Adding up the area of the bottom observations up to 5% of the total number of farms and expressing it as a percentage of total area gives (12).

Sources: IBGE in Souza Filho (1990) and IBGE (1985)

Table 3.5
Population of Espírito Santo: 1960-91

	Urban	%	Rural	%	Total	%
1960	378,744	29.2	919,498	70.8	1,298,242	100
1970	721,916	45.1	877,417	54.9	1,599,333	100
1980	1,293,378	63.9	729,962	36.1	2,023,340	100
1991	1,924,588	74.0	676,030	26.0	2,600,618	100

Sources: IBGE in Souza Filho (1990) and IBGE (1991)

Moreover, inequity in land distribution was aggravated (indicators 9, 10, 11, and 12) and rural-urban migration was intensified (Table 3.5). During the first half of the 1980s, the diffusion rate started to slow, mainly due to rural credit constraints and declining output prices; use of pesticides, for instance, diminished.

Main activities and related programmes

The GreenRevolution can also be comprehended by observing the changes in the main activities and related programmes.

Coffee After the 1960s Eradication Programmes, the government formulated plans to renew plantations, and these were embraced by farmers in the middle of the 1970s, when prices increased exceptionally due to frost in South Brazil. Domestic and external demand for *robusta* beans, which are blended with *arabica* beans, were expanding during the period. The plans provided conditions to promote better agricultural zoning and diffusion of chemical and mechanical inputs. Between 1975 and 1980, coffee production increased by 23.5% per year in the State (Table 3.2). During the first half of the 1970s, the percentage of Espírito Santo's coffee in Brazil's total production was around 5%; in the early 1980s this proportion increased to 13%. During the first half of the 1980s, although prices were declining (Figure 3.1), production continued increasing and the demand for credit was still high (Table 3.1). In the middle of the decade, a severe drought in Brazil pushed prices up. However, excess supply in the following years re-established the declining trend.

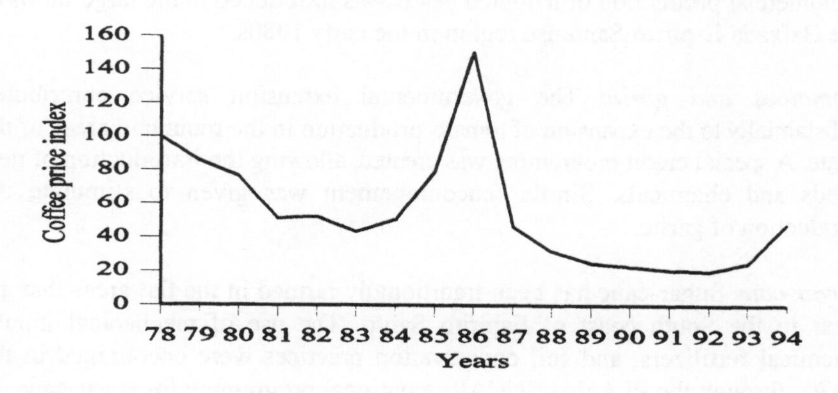

Figure 3.1. Coffee price index (Prices Received by Farmers, deflated by the IGP-DI, 1978=100)

Source: Fundação Getúlio Vargas (1994 and 1995)

Prices became further depressed after 1989, when the International Coffee Agreement's quota system broke down (de Melo, 1994). Profitability in coffee production at the end of 1980s and early 1990s was extremely low and many farmers decided to diversify into bananas, maize and cattle grazing. In 1994, prices increased due to frost in South Brazil.

Rice, beans, and maize Basic foodstuff production declined in the 1970s. Although there was rural credit provision, the local market was depressed and no special programmes were created to promote technological change in these activities during the decade. In the 1980s, however, urban markets sharplyincreased, particularly in Vitória, where large industrial plants were under construction. In addition, the Minimum Price Guarantee Policy (chapter 2) helped to encourage beans and rice production in the State. The latter was further stimulated by the PROVÁRZEAS (National Programme for Wetlands Drainage), which provided financial resources to bring 47 thousands hectares of wetland in Espírito Santo into agricultural production (Ministério da Agricultura, 1988). Rice production increased at 11.5% per year in the first half of the 1980s, mainly in the region of Colatina, where the programme was particularly successful.

Beans and maize had always been farmed as a subsistence crop, generally in association with coffee. Maize production increased during the 1960s as a substitute for coffee, and, in the 1970s, the extension service introduced new varieties of hybrid seeds and agri-chemical use. In the second half of the 1970s, maize farming was also favoured by increasing local poultry production. Commercial production of irrigated beans was introduced in the large farms of the Baixada Espírito-Santense region in the early 1980s.

Tomatoes and garlic The governmental extension service contributed substantially to the expansion of tomato production in the mountain areas of the State. A special credit programme was created, allowing the introduction of new seeds and chemicals. Similar encouragement was given to stimulate the production of garlic.

Sugar-cane Sugar-cane has been traditionally farmed in the flat areas that are next to the South coast of Espírito Santo. The use of mechanical inputs, chemical fertilizers, and soil conservation practices were encouraged in the 1970s through the PLANALSUCAR, a national programme for sugar-cane. In the early 1980s, the State of Espírito Santo was integrated into the second phase of the Proalcohol (National Programme for Alcohol), when seven new industrial distilleries and new plantations were established in the North.

Cocoa Cocoa farmers were assisted by the CEPLAC (Executive Commission for the Cocoa Plantations Plan), which was created in 1957 to provide technological and financial support to the Brazilian plantations. Production and productivity increased during the second half of the 1970s, when CEPLAC launched the PROCACAU (Programme for the Cocoa Plantations Recovering). In the following decade, however, investment in productivity augmentation decreased due to declining output price and credit constraints.

Black-pepper and papaya The production of black-pepper and papaya sharply increased during the first half of the 1980s. Favourable prices, good climatic conditions, and special financial support from the Bank of Espírito Santo for Economic Development created conditions to expand these enterprises in the North of the State.

Cattle grazing This activity increased from the early 1950s until the middle 1970s. In the North, large areas of the rainforest were cleared to allowed expansion of beef cattle grazing, while in the South grazing land for dairy farming expanded on depleted soils. Between 1967 and 1975, the World Bank provided credit to develop cattle grazing in Brazil. In Espírito Santo, these resources were allocated to improve pasture quality and animal health, introduce new breeds, and provide fodder during drought seasons. Beef processing industries were established in the North, while in the South the system for dairy production was further developed.

Planted forests Between 1960 and 1985, the area of homogeneous planted forests increased from 25 thousands to 158 thousands hectares. Most of these were subsidized eucalyptus plantations, which were owned by two large companies, Aracruz Florestal and Vale do Rio Doce. They would provide wood for cellulose and charcoal industries.

The environmental problems of the Green Revolution

Agricultural growth has caused negative effects on natural environment and human health in the State of Espírito Santo.[5]

Deforestation

The area of natural forests decreased from 1.36 million hectares in 1958 to 0.40 million hectares (8.3% of the State's territory) in 1990 (Fundação SOS Mata

Atlântica, 1993). The largest preserved forests are found in protected areas, such as the Caparaó National Park (26,000 ha), the Sooretama Biologic Reserve (24,000 ha), and the Linhares Florestal (22,000 ha), or in extremely hilly regions, where agriculture is difficult.

Soil depletion

1 Mechanical and intensive tillage have reduced organic matter and exposed the clay layer in drained wetlands of the mountainous region, resulting in soil compaction problems.

2 After indiscriminate and rapid deforestation, coffee has been farmed without soil conservation practices, causing serious problems in hilly areas: soil impoverishment, sedimentation in water courses, destruction of roads, floods, drought, and water contamination.

3 In plateau areas, use of conventional practices, especially heavy mechanization, have reduced the organic matter layer and caused soil compaction.

4 In areas near the Atlantic coast, sandy soils became extremely vulnerable after native vegetation was cleared. Use of fire and mechanization have further reduced the already poor availability of organic matter.

5 In the North of the State, there are large areas of flooded turf soils, which have been drained to allow rice farming and cattle grazing. Given the high concentration of organic matter and extreme acidity of these soils, farming requires a high dosage of correctives. Drainage and subsequent agricultural use have reduced the organic layer by both accelerated decomposition in aerobic conditions and use of heavy machinery.

Water: pollution and management

1 Residues of chemical fertilizers and pesticides from horticulture have been found in the Santa Maria River and some of its tributaries. Fish stocks have been reduced, and use of contaminated water has created health problems to rural communities (Reis, 1987). Other rivers and streams have suffered similar problems, especially due to intensive use of agri-chemicals in coffee plantations.

2 Contamination of underground water by nitrates has been detected in the regions of Santa Maria and Fartura, where use of agri-chemicals in horticulture is intensive.

3 In the driest areas of the North, irrigation has reduced water availability for domestic use.

4 Drainage of wetlands for agricultural purpose has also reduced water availability.

5 In several areas, superficial water run-off speed has increased due to both deforestation and conventional agriculture's lack of green cover. Consequently, underground water recharging and water availability during dry seasons have been reduced. In recent years, rural communities and small cities, such as Presidente Kennedy and Conceição da Barra, have faced seasonal shortages due to this problem.

6 Deforestation and inadequate farming systems have exacerbated soil erosion. Increasing sedimentation on the riverbeds has caused frequent floods, with harmful effects on agriculture and riverine communities. The estuary fishery has also been damaged.

Rural workers' health

Indiscriminate use of agri-chemicals, inadequate disposal of containers, and lack of protective clothing have caused health problems in the horticultural region of the Santa Maria de Jetibá. Cases of human poisoning have been recorded (Reis, 1987).

Conclusion

The recent history of Espírito Santo's agriculture can be divided into two distinct phases. The abolition of slavery in 1888 marks the start of the first, which finished in the 1960s with the introduction of the Coffee Erradication Programmes. During this phase, Espírito Santo's agriculture had a subsistence, family labour-based structure. Coffee was the main, and sometimes the only, commercial output. On-farm production of basic foodstuffs ensured households' survival, and low-external-input coffee farming provided a monetary surplus. This structure was strongly resistant to coffee price crises, although its long-term sustainability was questionable, given that it increased extensively by clearing the rainforest and destroying soil fertility.

The second phase started in the early 1970s and was characterised by the Green Revolution. A new structure, based on technological advance, entrepreneurial administration of the farm, and the large use of hired labour and external inputs, was created. Although production and productivity increased, farmers became more vulnerable to market changes. The use of family labour decreased, while that of seasonal labour increased. Concentration of land property and rural-urban migration were accelerated. The introduction of the new technologies created ecological imbalances and aggravated environmental problems. One can conclude that this growth pattern was economically, socially, and ecologically unsustainable.

The number of farmers adopting more sustainable agricultural practices in Espírito Santo has increased since the second half of the 1980s. The reasons for this are examined in chapter 8. It is hypothesised that not only non-economic considerations, such as environmental and health concern, but also farmers' economic constraints in dealing with Green Revolution technologies have stimulated adoption of new approaches to agricultural production in the State.

Notes

1. For a detailed account of the economic history of the State of Espírito Santo see Rocha and Cossetti (1983).
2. The plantations were extremely destructive. After the clearing of the rainforest and a few years of farming, soil fertility would be exhausted. Given that land was abundant, production could grow for many years, reaching the borders of the Provinces of Espirito Santo and Minas Gerais and beyond. By the end of the nineteenth century, coffee plantations were introduced in the plateaux of São Paulo, providing the economic basis for the Brazilian process of industrialization (Cano, 1985).
3. The slavery system was abolished by a government law in 1888.
4. For more detailed examination of the Green Revolution in Espírito Santo see Souza Filho (1990).
5. For a report on the environmental problems of the State of Espírito Santo see Comissão Coordenadora do Relatório Estadual sobre Meio Ambiente e Desenvolvimento (1991).

4 Sustainable development and agricultural sustainability

Introduction

The notion of economic development has been changing in recent years towards the more consensual view that environmental conservation and better standards of living must be pursued simultaneously. Several attempts have been made to connect issues such as economic growth, exploitation of natural resources, concern with future generations, quality of life, income distribution and poverty. Most of these themes had been neglected, or scantily treated, by economists in the past. The dissemination of the idea of sustainable development however has put them in the core of economic development thought.

This chapter provides an overview of several issues related to the concept of sustainable development, which covers a very broad range of subjects and has sometimes given rise to extremely imprecise definitions and conclusions.[1] There is no attempt to elaborate new concepts. However, a review of some of the different themes could reveal aspects which can make adoption of sustainable agricultural technologies a workable issue.

The concept and measurement of economic development: an old debate

In the early 1960s Okun and Richardson (1965:230) defined economic development as 'a sustained, secular improvement in material well-being, which we may consider to be reflected in an increasing flow of goods and services'. This definition strongly referred to the material plane of development, although the authors expressed concern with respect to other aspects of the term. In the following quotation they recognise that material well-being and social life at large may not necessarily follow the same direction; economic progress (growth) does not imply development in a broader sense:

> Economists realize that it is a long step from material progress narrowly construed to improvements in social welfare in the broadest sense of the term. Generally, we expect economic progress to contribute positively to social well-being. But it is perfectly conceivable that social improvement may not always be coincident with economic progress; such progress may even occur at the expense of a deterioration of other, highly valued aspects of life (Okun and Richardson, 1965, p.230).

Once the limits of the concept have been considered, a measure of economic development could, in principle, be taken in terms of the flow of goods and services. A natural candidate is the Gross National Product (GNP). The authors point out this indicator has some *known* weaknesses. It includes the value of investment goods that are mere replacements for the old capital stock and do not contribute to increased economic capacity.[2] Eliminating depreciation from GNP results in the Net National Product (NNP). However, it is an aggregate measure and cannot be taken as an index for the individual economic welfare. A growth in the aggregate national income can be accompanied by a more rapid growth in the population. Real per capita income would be more appropriate, but it is an average; the distribution of income, which is critical in assessing economic welfare, does not enter the calculation.

Another problem in measuring economic development, which is also identified in an old conceptualization of economic development such as in Okun and Richardson (1965), refers to the depletion of natural and non-renewable resources:

> One can argue that depletion of such resources is not relevant to any measure of economic well-being or welfare if the present population of a country chooses to disregard the future entirely. If the population does place some value on the conservation of resources for future generations, however, the depletion of natural resources should perhaps be regarded as a negative factor in any assessment of economic welfare and the economy's performance (Okun and Richardson, 1965, p.233).

These problems in measuring economic development were pointed out long before the concept of sustainable development had been created and disseminated among economists and environmentalists around the world. As will be seen in this chapter, the concept of sustainable development is intended to overcome these difficulties.

The limits to growth

The Club of Rome's attempt at modelling long-term, global problems resulted in the publication of Limits to Growth (Meadows et al., 1972). Five major trends of global concern were investigated by the study: accelerating industrialization, rapid population growth, widespread malnutrition, depletion of non-renewable resources, and deteriorating environment. Having detected an exponential increase in human activities, it was realized that this trend would change in a possibly auto-destructive way when it approached the 'carrying capacity' of the planet. Population, for instance, cannot exponentially increase forever, because further increase beyond a certain stage is constrained by the environment.

The optimistic view that technology can always be applied to alleviate environmental pressures is challenged in Limits to Growth. Although technological development is considered vital to the future of human society, the application of technology to problems of depletion of resources, pollution, and food shortage does not solve the essential problem, which is exponential growth in a finite system. Technological progress can delay the system's collapse, but it is not able itself to restrain growth of either population or capital stock. Furthermore, technological change can bring undesirable 'social side-effects'. The Green Revolution, for instance, has achieved important results in terms of increased food production, but in several places it has aggravated problems related to income distribution, migration to urban areas, agricultural unemployment, and malnutrition.

The Club of Rome's conclusion was that deliberate constraints should be imposed on growth to achieve equilibrium of constant population and capital stock. In this steady-state system, the levels of capital and population should be set according to society's values, allowing for revisions and adjustments determined by technological advances. Technology is fundamental in the sense that it can contribute by avoiding nonrenewable resource shortages, reducing pollution and changing agricultural and industrial activities towards more conservation-oriented systems. Moreover society must shift its preference more towards services such as education and health facilities and less towards material goods. To alleviate problems related to income inequalities, capital should be diverted to agriculture to maintain everyone at least at subsistence level. Resources would still be gradually depleted, but at a pace slow enough for technology and industry to adjust.

The WCED's concept of sustainable development

In 1982 the United Nations established the World Commission on Environment and Development to address these issues. The work of the Commission resulted in the 1987 Brundtland report, which was perhaps the major impetus for the dissemination of the concept of sustainable development.

The historical process that led to the Brundtland report can be traced back to the 1972 UN Conference on the Human Environment held in Stockholm. This Conference was initially motivated by developed countries' concern with the environment. Developing countries however tried to shift the emphasis towards other development issues.[3] For them, alleviation of poverty was more important than improvement or preservation of the environment, which was considered a luxury they could not afford.[4] Nevertheless, the landmark of the Conference was the recognition of the interaction between environment and development.

After the 1972 Conference, several initiatives on international environment and development were held: Law of the Sea, London Dumping Convention, Basel Convention, Vienna Convention on the Protection of the Ozone Layer (and its Montreal Protocol), regional seas programmes, regional agreements on air pollution, and the Brandt Commission on North-South issues (Grubb et al., 1993). Despite these attempts, environmental problems continued to increase as a consequence of the social and economic development patterns followed by the world. By the early 1980s, progress since Stockholm was not auspicious and this led the United Nations to establish the WCED. Increasing environmental concern in developed countries and social and economic problems in developing countries led the Commission to coin the term 'sustainable development'.

> Sustainable development is development that meets the needs of the present without compromising the ability of future generations to meet their own needs. It contains within it two key concepts:
> - the concept of 'needs', in particular the essential needs of the world's poor, to which overriding priority should be given; and
> - the idea of limitations imposed by the state of technology and social organization on the environment's ability to meet present and future needs (WCED, 1987, p.43).

The concept tries to balance both sides' problems, allowing for each to emphasise its own necessities. To reach sustainable development, developed countries should pursue policies such as recycling, energy efficiency, conservation, rehabilitation of damaged landscapes, while developing nations should seek equity, fairness, respect for law, redistribution of wealth and wealth creation (Sandbrook, 1992).

During the 1970s and 1980s, several issues such as the oil crisis, the hole in the ozone layer, extinction of some species, and increased public concern with the environment, showed that problems related to the depletion of natural and nonrenewable resources should not be marginalised in development studies. The idea that natural resources are limited should be viewed from a broader perspective. In the WCED's concept of sustainable development it occupies a central position. The concept emphasises the problem of income distribution by not only recovering the notion of justice between individuals in the same generation, but also by broadening it to incorporate the idea of intergenerational equity. In this sense, conservation became a matter of temporal distribution of welfare.

The environment and economic growth: tradeoff or complementarity

Economic growth is a key element in the discussion of development. Even before the environmental debate had reached economics, economists knew that it was a long step from material progress to improvements in social welfare. As the environmental debate progressed, mainly after the publication of the Club of Rome's Limits to Growth , the idea that 'in general' economic growth leads to improvement in welfare was challenged. It was suggested that there was a tradeoff between economic growth - measured by rising real per capita incomes - and environmental quality. This was a radical shift from the prevailing view, even considering that the negative effect of the depletion of natural and nonrenewable resources on economic welfare had been pointed out before. The idea of zero economic growth as a necessary condition for the maintenance of environmental quality was not readily acceptable, particularly by developing countries. In the 1972 UN Conference on the Environment they posited the view that economic growth would be a necessary condition for alleviating poverty. Considering the need to conciliate the views, the concept of sustainable development suggested by WCED (1987) admitted that in places where essential needs have not been met, there is a complementarity between economic growth and improvement in environmental quality.

> Poverty reduces people's capacity to use resources in a sustainable manner; it intensifies pressure on the environment ... A necessary but not a sufficient condition for the elimination of absolute poverty is a relatively rapid rise in per capita incomes in the Third World ... Growth must be revived in developing countries because that is where the links between economic

growth, alleviation of poverty, and environmental conditions operate most directly (WCED, 1987, p.49).

Developing countries' organizations, such as the Economic Commission for Latin America and the Caribbean, endorsed that view:

> Poor people live in areas where natural resources are scarce and the environment has greatly deteriorated ... this deterioration is the result of displacement of their activities to areas where natural capital is not very highly regarded (having minimum available and obtainable rent) or where other forms of capital are virtually absent. This displacement leads to a vicious circle of poverty (destroy and survive). The lower incomes are, the more short-term oriented consumer choices will become because immediate need ... poverty will not be eradicated unless poor people are given a better chance to accumulate capital (ECLAC, 1991, p.73-74).

A conciliatory view is given by Pearce and Turner (1990). For them, growth in the standard of living occurs as the environmental capital is developed and expanded. However, this process has a limit, where further increase in the standard of living would reduce the environmental capital, unless an additional mechanism, such as technical progress, is enforced.

Neoclassical economics and the environment

Neoclassical economics has incorporated environmental issues at both the micro and macro levels. In microeconomics, the theory of externalities identifies environmental problems as a case of 'market failure'. Many services provided by natural environments have zero price because there is no market in which their true economic value can be revealed (Pearce et al., 1994). Possible solutions are either to attach a quantitative weight to human desires and preferences (e.g. contingent valuation) or to consider the cost of deterioration in the price system (e.g. Polluter Pays Principle).

In macroeconomics, the best example is given by Hartwick's (1977) and Solow's (1986) models for the intergenerational allocation of exhaustible resources. In Hartwick's model, an economy, which is represented by a macro production function, is assumed to produce current output with the use of a stock of capital accumulated from previous investment, and withdraws from a given stock of a nonrenewable resource. The model does not allow for technical progress, assumes constant costs of production and full capital and labour employment. The main conclusion is: if society uses a nonrenewable resource so that its price increases at the rate of interest (Hotelling's rule), and invests in reproducible capital goods the competitive rents obtained from the use of that resource (Hartwick's rule), then

> this society will find that it is just able to maintain a constant stream of consumption. The accumulation of reproducible capital exactly offsets the inevitable and efficient decline in the flow of resource inputs (Solow, 1986, pp.144-145).

The incorporation of the environment into neoclassical thought raises the question of valuation, which is not new in the field of economics. Valuing the environment, or valuing the environment as a stock of capital, brings back some aspects of the capital theory and its famous controversies (Harcourt, 1991). The use of a production function to determine future long term equilibrium means that a valuation method for the input variables has to be adopted. Valuing capital by its past cost means projecting the past into an uncertain future or, in other words, it implies that nothing but known events will happen. In the case of environmental goods, further complications arise because cost of production has not been recorded in an economic sense. It is possible to value a capital good as a sum of the discounted stream of future profit which it will earn. In fact, competitive markets should reveal this value through current prices. However, 'that value is clouded by the uncertainty that hangs over the future' (Robinson, 1956, p.117). When an unexpected event occurs all variables (prices, physical productivity, rate of discount) are subject to change. Thus, any attempt to predict long term equilibrium through a production function can only be possible by assuming perfect knowledge of all relevant future prices. Even accepting this possibility, there remains a methodological problem; future prices are assumed as given when the purpose of the production function is to determine prices (Robinson, 1954).

Hartwick/Solow's model, and any attempt to evaluate the environment as a stock of capital, is hampered by the problems pointed out by the capital controversy. It has also two other additional drawbacks. First, the elasticity of substitution between natural and reproducible resource is assumed to be one. As there are some services from the natural resources that man-made capital cannot provide (e.g. protection from solar radiation by the ozone layer), there is no reason to assume a unitary elasticity between the two. Second, the system is not sustainable in a biophysical sense.

Within Solow/Hartwick formulation, the constraint on net investment is a

> financial one, whereas there should be a physical one, in that, depreciation of the physical stock has to be compensated for by extraction of the exhaustible resources. Over time, the model suggests growing capital stocks, implying higher depreciation to be met from declining physical stocks of exhaustible resource (Young and Burton, 1992, p.16).

Another criticism of the neoclassical attempt to value the environment is given by Daly and Cobb's (1989) rejection of the discount rate. Future satisfactions/dissatisfactions derived from preserving/depleting the environment are discounted and added up to provide the present value. As future generations cannot manifest their desires, this task is fulfilled by the present generation, although it will not be alive to experience these satisfactions/ dissatisfactions in the future.

> Discounting present value represents the value to present people derived from contemplating the welfare of future people. It does not reflect the welfare of future people themselves, or even our estimate of their welfare. Rather it reflects how much we care about future people compared to ourselves (Daly and Cobb, 1989, p.154).

Discounting satisfaction/dissatisfaction (accepting that it is known) of future generations has implications for the intergenerational distribution issue. If egalitarian treatment is to be given, a positive rate of discount would only be acceptable under condition of growing productivity. However, as Daly and Cobb (1989) argue, if the productivity calculation included the cost of the technological risk (radiation, toxic waste, accidents) brought about by the technologies created to compensate environmental losses (depletion of high-quality resources, pollution and habitat destruction), it would reveal a decline rather than an increase. Then the rate of discount should be negative in order to provide an egalitarian distribution of welfare.

The problem of valuation of environmental capital can be solved by measuring it in terms of physical stocks. As natural resources are highly diversified, this would result in serious problems of aggregation, which prohibit the application of the method to neoclassical models. In a practical sense however it has great appeal. As the issue of intergenerational distribution is surrounded by uncertainty about the future, the adoption of 'safe minimum standards' (Bishop, 1978), in which indicators of physical stocks would be used to allow for critical minimum reserves, could be a feasible solution.

Alternative views on sustainable development

Although neoclassical thought has great appeal in a world in which the basis for policy intervention is dominated by market liberalism, alternative views on development and environmental issues have been expressed.

Daly's biophysical and ethicosocial limits to growth

For Daly (1987), growth and development must be properly differentiated:

> By growth I mean quantitative increase in the scale of the physical dimensions of the economy; i.e., the rate of flow of matter and energy through the economy (from the environment as raw material and back to the environment as waste), and the stock of human bodies and artifacts. By development I mean the qualitative improvement in the structure, design, and composition of physical stocks and flows, that results from greater knowledge, both of technique and of purpose (Daly, 1987, p.323).

In this sense, development is not necessarily limited but growth presents two classes of limit: the biophysical and the ethicosocial. The biophysical limit is determined by the second law of thermodynamics (the entropy law), which prevents complete recycling. Growth means transformation of the low entropy of the planet into high entropy, which is an irreversible physical flow. Given that the planet's stock of low-entropy materials is limited, only for a limited period can the physical dimensions of wealth and income grow. For most low-entropy sources, upon which modern civilization is based, depletion may reach high levels in the short run.

Daly presents four ethicosocial arguments that limit growth. The first refers to the intergenerational distribution of nonrenewable resources. Any temporary economic expansion can only be achieved by depleting 'geological or ecological capital', which should be considered as an unacceptable cost imposed on future generations. The present generation has the 'moral obligation' to leave a certain endowment for future generations. Second, economic growth requires increasing space for material goods and people, which means the takeover of the habitat and consequent extinction of another species. This takeover has not only the limit imposed by the breakdown in 'ecological complexity and interdependency' but also the limit imposed by the 'moral obligation' to concede subhuman species the 'right to experience pleasure and pain'. A third ethicosocial limit to growth is determined by self-cancelling effects on welfare. Happiness is assumed to be related to relative differences in interpersonal income. In this sense, growth that increases everyone's income but maintains relative positions is not important for human welfare. Consequently, for societies in which the level of 'absolute needs' has been met, other goals but growth should rise relatively in the scale of social priorities. Fourth, the glorification of self-interest and the scientific-technocratic view, which are attitudes that foster growth, lead to the erosion of values like honesty, sobriety or trust that are the basis of social cohesion. As such moral controls, which are necessary to a market society are eroded, then external police power is created. The latter requires further depletion of resources if it is to compensate for the depletion of the moral restraints.

Coevolutionary agricultural development

Norgaard (1984, p.531) uses coevolutionary theory to provide a link between ecology and economics:

> Coevolution in biology refers to an evolutionary process based on reciprocal responses of two closely interacting species ... The concept can be broadened to encompass any feedback process between two evolving systems. For agricultural social and ecological systems, man's activities modify the ecosystem while the ecosystem's responses provide cause for individual action and social organization. Thus, agricultural development can be viewed as a coevolution process between a sociosystem and an ecosystem that, fortuitously or by design, benefits man.

Norgaard points out that a 'coevolutionary process' merely refers to the reciprocal process of change. It is different from 'coevolutionary development' in the sense that it is not necessarily positive for man. The Green Revolution in Western Europe and North America is an example of coevolution that cannot be characterized as a coevolutionary development. It coevolved from a small-scale, labour-intensive, near subsistence farming system to a large, mechanized, energy-intensive, monocultural commercial farming system in which short-run ecological and financial stability are guaranteed by the use of agrichemicals and the creation of risk-spreading institutions. Technological advances and institutional arrangements provided conditions for transforming monoculture into an attractive option for farmers. However, as in a coevolutionary process, the ecosystem reacted by creating resistance to pesticides. Then, new pesticides were developed. As the level of contamination increased, regulatory bodies were added to the already large set of institutions that maintain the feasibility of the system. In this process, a large quantity of energy and human capital is always exerted in the same direction. There are enormous costs and no beneficial changes. The dynamic interaction between ecosystem responses and sociosystem responses offers a catastrophic long-run coevolutionary perspective.

The distinction between the coevolutionary process and coevolutionary development has implications for the optimal exploitation of natural resources. The following quotation suggests the basis on which the coevolutionary approach addresses this question:

> Coevolutionary agricultural development can be envisioned as a sequential process in which a surplus of energy and human capital, beyond what is necessary to maintain the ecosystem and sociosystem in their present states, is directed to establish a new interaction between the systems. If this new interaction is more favourable to man and a surplus can be directed to further beneficial changes, then coevolutionary development is underway (Norgaard, 1984, p.351).

The coevolution of Western agriculture is depleting nonrenewable resources in a way that is not affecting coevolutionary development. These resources should be used to change the direction of the system's (social and ecological) interactions towards long term sustainability, for instance, by the development of nitrogen-fixing plants. Thus, coevolutionary agricultural development conforms with Daly's concept of development in that both place emphasis on structural/qualitative changes to maintain a favourable long-run trajectory. Perhaps the stimulus to take this direction may occur when the stock of nonrenewable resources is close to complete depletion. This path however may limit the possibilities for future beneficial coevolution (Young and Burton, 1992).

Agricultural sustainability and resilience

Conway (1987) and Conway and Barbier (1990) used the idea of resilience to define sustainability at the agrosystem level. Sustainability of an agrosystem is determined by its ability to maintain productivity in the face of disturbing forces. Two types of forces may affect a system, stress or shock. The effects of stress are small in the short run, though its cumulative impact can predictably be large in the long run. Examples are erosion, salination, or declining output prices. Shock is an unpredictable but transitory change, such as a new pest, drought, flooding, or significant increase in input prices, due to, say, an oil crisis.

For Conway and Barbier there are four criteria by which agricultural development can be judged: sustainability (as defined above), productivity, stability, and equitability. Productivity is defined as the output of valued product per unit of resource input, while stability is the constancy of this productivity in the face of small disturbing forces such as climate. Equitability refers to the fairness of distribution of the productivity among those involved in the agricultural system. There are complex tradeoffs between them. For example, although monoculture systems can increase productivity, the necessary use of chemicals and machinery compromise sustainability. The role of agricultural research is to account for these tradeoffs and develop systems that maintain or improve productivity without loss of sustainability. In some areas of developing countries, for instance, where Green Revolution packages negatively affected equitability, it is particularly important to identify new production systems in which equity and sustainability can be improved without compromising productivity.

Tisdell (1988) points out that Conway's approach to sustainability has similarities with the idea of stable equilibrium. A system can be said to be in a state of stable equilibrium when it is able to return to its original path after the impact of any shock. This condition however is only fortuitously reached, since after a shock the system parameters may change in a way that makes it impossible to return to the original equilibrium. Either a new equilibrium or an unstable disequilibrium process may be achieved. These possible subsequent states cannot be judged a priori, given that a society's values could change during these processes.

The above criticism has a similarity with the coevolutionary approach in the sense that any return to an original equilibrium state should be understood as a mere result of the interactions between the ecosystem and sociosystem.[5] In other words, recovering an original state is only one possibility among others, which may also be considered 'good for man'. The ecosystem and the sociosystem do not need to be in a state of equilibrium. They could be continuously changing 'to the benefit of man'. This raises the question of what does benefit man. The answer is subject to individual ideals and concepts of what a good society is. In terms of agricultural development, Conway and Barbier, on the basis of their experience in less developed countries, found that the four criteria (productivity, sustainability, stability and equitability) offered a practical way to judge the matter. Because there are tradeoffs, the best balance is still a society decision.

Sustainable agricultural technologies

The debate about how to achieve sustainability in agriculture is troubled by disputes and disagreements over which elements of production are acceptable and which are not. There are several different agricultural technologies which are classified in the literature as sustainable, although the sustainability of farms where they are employed may be put in question by the advocates of one or another line of thought (e.g. organic monoculture). As pointed out by Ikerd (1993, p.31),

> Some contend that sustainability must be achieved by fine tuning conventional systems of farming. They do not believe that lower-input or organic systems of farming will ever be capable of feeding the growing world population. Others argue that sustainability will require a different model or paradigm for farming that relies less on commercial inputs and more on farm resource management. They see the input-dependent industrial model of agriculture as being fundamentally incompatible with maintaining a healthy ecological and social environment. Advocates of organic farming believe that sustainability will require the total elimination of manufactured chemical inputs. Others propose still different models of farming as a means of achieving long run agricultural sustainability.

OECD (1994) has stressed the difficulty of imposing a rigid definition of sustainable agriculture in the face of the enormous variety of social, economic and environmental contexts that characterise countries and even regions in the same country. However, it is possible to obtain a consensus that

> ... sustainable forms of agriculture are characterized by the adoption of practices and technologies that:
> - use integrated management techniques which maintain ecological integrity both on and off the farm;
> - are necessarily site-specific and flexible;
> - preserve biodiversity, landscape amenity and other public goods not valued by existing markets;
> - are profitable to producers in the long-term; and
> - are economically efficient from a societal perspective (OECD, 1994, p.8).

There are several terms related to sustainable agricultural technologies. While most of these terms refer to specific farming practices or systems, others, such as 'alternative agriculture' and 'LEISA' (low-external-input and sustainable agriculture), have a broader meaning. Table 4.1 presents definitions of some farming system which are quoted in the literature as sustainable.

The term 'alternative agriculture' is one of great generality. Its most comprehensive definition is given by the National Research Council (1989, p.3):

Table 4.1
Definitions of some farming systems with great potential for sustainability

Biodynamic farming A holistic system of agriculture devised by Rudolph Steiner that seeks to connect nature with cosmic creative forces. An attempt is made to create a whole-farm organism in harmony with its habitat. Compost and special preparations (e.g. plant derived sprays) are used. Synthetic fertilizers and pesticides are avoided.

Ecological agriculture Farming practices that enhance or, at least, do not harm the environment and are aimed at minimising the use of chemical inputs, rather than completely avoiding them as in organic farming. Also known as ecofarming.

Natural farming A system of agriculture devised by Masanobu Fukuoka that seeks to follow nature by minimising human interference: no mechanical cultivation, no synthetic fertilisers or prepared compost, no weeding by tillage or herbicides, no dependence on chemicals.

Organic farming A system of agriculture that encourages healthy soils and crops through practices such as nutrient recycling of organic matter (such as compost and crop residue), crop rotations, proper tillage and the avoidance of synthetic fertilisers and herbicides.

Permaculture A consciously designed, integrated system of perennial or self-perpetuating species of crops, trees and animals.

Source: Reijntjes et al. (1992)

> In contrast to conventional farming ... alternative systems more deliberately integrate and take advantage of naturally occurring beneficial interactions. Alternative systems emphasize management; biological relationships, such as those between the pest and predator; and natural process, such as nitrogen fixation instead of chemically intensive methods. The objective is to sustain and enhance rather than reduce and simplify the biological interactions on which production agriculture depends, thereby reducing the harmful off-farm effects of production practices.

Here alternative agriculture is not defined as a single practice or system, but any farm system that pursues the above objective. They are often diversified and tend to increase stability and resilience, and reduce financial risk. According to the National Research Council, examples of alternative farming systems are those known as biological, low-input, organic, regenerative, or sustainable. The practices and principles they emphasize are:

- Crop rotations that mitigate weed, disease, insect, and other pest problems; increase available soil nitrogen and reduce the need for purchased fertilizers; and, in conjunction with conservation tillage practices, reduce soil erosion.
- IPM (Integrated Pest Management), which reduces the need for pesticides by crop rotations, scouting, weather monitoring, use of resistant cultivars, timing of planting, and biological pest control.
- Management systems to control weeds and improve plant health and the abilities of crops to resist insect pests and diseases.
- Soil- and water-conserving tillage.
- Animal production systems that emphasize disease prevention through health maintenance, thereby reducing the need for antibiotics.
- Genetic improvement of crops to resist insect pests and diseases and to use nutrients more effectively (National Research Council, 1989, p.4).

Another term which is found in the literature on sustainable agricultural technology refers to the set of practices known as LEISA. The following definition is given by Reijntjes et. al. (1992, p.XVIII):

> LEISA is agriculture which makes optimal use of locally available natural and human resources (such as soil, water, vegetation, local plants and animals, and human labour, knowledge and skills) and which is economically feasible, ecologically sound, culturally adapted and socially just. The use of external inputs is not excluded but is seen as complementary to the use of local resources and has to meet the above mentioned criteria.

Advocates of this approach claim that LEISA is able to supply at least the basic human needs while maintaining or enhancing the quality of the environment and conserving natural resources. An efficient use of local resources can solve financial and ecological problems of farmers who are not in a position to use artificial input or can only use small quantities. Most of these farmers may be practising erosive forms of low-external-input agriculture such as exploiting land beyond its carrying capacity, and deforestation. There are several reasons for adopting this kind of farming system: input becomes more expensive (e.g. due to the debt crisis), output prices decrease, sustainable technologies are not known, farmers may be moving towards marginal lands, the commercial infrastructure (transport, distribution of input, financial institutions, and intermediaries) is not adequate, there is lack of property rights, etc. Table 4.2 shows some examples of LEISA practices.

Table 4.2
Examples of LEISA technologies

Composting Composting is the breakdown of organic material by micro-organisms and soil fauna to give a humus end product called compost. It is an important technique for recycling organic waste from postharvest processing, dung, nightsoil, urine, etc.) and for improving the quality and quantity of organic fertilizer.

Green manure Trees, shrubs, cover crops, grain legumes, grasses, weeds, ferns and algae provide green manure, an inexpensive source of organic matter and fertility.

Mineral fertilizer Mineral fertilizer normally increases the availability of biomass for organic fertilizer and may enhance soil life when applied moderately.

Mulching Mulch can be defined as a shallow layer at the soil/air interface in which the most traditional composition includes dry grass; crop residuals (straw, leaves, etc.); fresh organic material from trees, bushes, grasses and weeds; household refuse and live plants (cover crops, green manures). It is an important technique for improving soil microclimate; enhancing soil life, structure and fertility; conserving soil moisture; reducing weed growth; preventing damage by impact from solar radiation and rainfall (erosion control); and reducing the need for tillage.

Intercropping The growing of two or more crops simultaneously on the same field has beneficial effects in terms of better control of insects, diseases and weeds.

Trap and decoy crops Various kind of traps can be made to catch insects, rodents or other creatures which threaten crops or livestock. The most common is the light trap, set up to catch night flying insects. Pests can also be attracted by certain plants. When these are sown in the field or alongside it, insects will gather on them and can thus be easily controlled.

Biological control In biological control, pests are supressed by their natural enemies, such as birds, spiders, miter, fungi, bacteria, viruses or plants (e.g. cover crops to control weeds).

Plant-derived pesticides Numerous plants have defensive or lethal effects on vertebrates, insects, mites, nematodes, fungi or bacteria. Active components can be extracted from various parts of plants and dispersed over the crop. Unfortunately this age-old knowledge is rapidly being lost, particularly where chemical pesticides have been introduced.

Integrated crop-livestock-fish farming Such systems are conducive to nature conservation as they promote habitat stability and diversity for the wildlife living on the farms and in adjacent areas. As these integrated systems optimise the use of on-farm and adjacent resources, they encourage habitat conservation rather than destruction. Such systems are productive and profitable because they utilise waste as inputs in other enterprises within the farm, and because fish are a higly nutritious and valuable traditional food. They use microenvironments within a farm system which add to farm productivity and security.

Minimum tillage Soil management practices which seek to minimise labour inputs and soil erosion, to maintain soil moisture and to reduce soil disturbance and exposure. Crop stubble is left or mulch is applied to protect soil. Also known as conservation tillage or reduced tillage. In its most extreme form (zero- or no-tillage), seeds are drilled directly into the otherwise undisturbed soil.

Multiple cropping Growing two or more crops in the same field in a year, at the same time, or one after the other, or a combination of both.

Multistorey cropping Growing tall crops (often perennials) and shorter crops (often biennials or annuals) simultaneously.

Source: Reijntjes et. al. (1992)

It is also claimed that farmers who practise high-external-input agriculture can also reduce contamination and cost, and increase the efficiency of external input by adopting some LEISA techniques. Here, the internal-external resources dichotomy cannot be taken as an absolute criterion to distinguish sustainable practices from unsustainable ones. Although the use of external input can sometimes be associated with pollution and depletion of nonrenewable resources, there is no reason for excluding hybrids, for instance, or mineral fertilizers if they can be integrated in a sustainable way (Young and Burton, 1992).

The development of LEISA practices can be carried out with the joint participation of farmers and professionals from outside their community (e.g. extensionists, researchers). This kind of approach, which connects LEISA to PTD (Participatory Technology Development), is described by Reijntjes et. al. (1992, p.XIX) as

> ... a process of creative interaction within rural communities, in which indigenous and scientific knowledge are combined in order to find solutions

to farmers' problems and to take the fullest possible advantage of local opportunities. It involves collaboration of farmers and development agent in analysing the local agroecological system, defining local problems and priorities, experimenting with various potential solutions, evaluating the results and communicating the findings to other farmers.

PTD is an alternative to the single commodity and station-based research that is practised by most official agricultural organizations. In fact, non-governmental organizations have been particularly active in its development, mainly in developing countries, where they have been promoting diffusion of sustainable agricultural technologies.

Conclusion

This chapter provides a review of several issues related to sustainable development. Many questions that are currently at the core of economic development thought have been brought to light thanks to the dissemination of the idea of sustainable development. It has been recognised that there are complex links between economic growth, environment, quality of life, concern for future generations, income distribution and poverty. Because sustainable development has this broad range of intricate issues, it is difficult to define. Some of its advocates suggest that sustainable development implies no growth, while others are in favour of qualified patterns of growth. Most of these conflicts derive from two difficulties. First, there is a large diversity of social, economic and environmental contexts, in which practical solutions to local problems require specific policies. Second, decisions that have long-run effects are surrounded by great uncertainty about future technologies and society's values. These uncertainties are the most important factor precluding precise sustainability criteria. Sustainable development is in fact a process in which certain directions have to be pursued and constantly re-assessed on the basis of a changing world.

In agriculture, there are conflicts about which technologies are sustainable and which are not. However, there is a consensus that both conservation and better standards of living must simultaneously be pursued in a long-run perspective. In a practical sense, sustainable farmers are not necessarily those who adopt a rigid set of practices regulated by a particular organization, but those who intentionally make a move towards this consensual goal.

Notes

1. There is a large number of alternative definitions of sustainable development. The annex of Pearce et al. (1994) identifies at least 24.
2. This view can be questioned because it ignores the dynamics of the technological change. When a firm replaces its machinery, generaly, the new capital incorporates innovations that improve the efficiency of the production system. The value of the old stock of capital is destroyed, but replaced by new production systems (see Schumpeter's (1939) 'creative destruction'). This process has implications for environmental conservation and welfare, as new equipment can incorporate environment-friendly innovations.
3. In 1971, as part of the preparatory activities for the 1972 United Nations Conference, the Economic Commission for Latin America and the Caribbean organized a meeting in which it was recognized that 'the low level of development of the countries contributed to the deterioration of the environment...' (ECLAD, 1991, p.17)
4. 'In order to tackle poverty, they were prepared to adopt Western ways and accept the environmental problems as part of the package' (Sandbrook, 1992, p.15).
5. Even Conway and Barbier (1990, p.37) recognize that 'sustainability is a function of the intrinsic characteristics of the system, of the nature and strength of the stresses and shocks to which it is subject, and of the human inputs which may be introduced to counter these stresses and shocks'.

5 Review of adoption and diffusion of technology theory

Introduction

The purpose of this chapter is to explore the main themes in the theory of adoption and diffusion of new technologies. Many theoretical developments and much applied work have been produced in recent decades, making a complete examination a huge task for this study. Our major objective is to provide a background to the methodological approach chosen for the empirical investigation of chapter 8. Some of the first empirical analyses of diffusion of innovations were applied to agricultural technologies. Most studies in this area were related to the production packages that came with the Green Revolution, for which the technologies we are interested in, sustainable practices, are substitutes. However, given that many of the theoretical developments were in other fields of economics, a broader picture may provide some insights, which may not appear in a restricted review.

Origin of innovations

Schumpeter (1939) defined invention as the development of a new artifact or process, while innovation represents its economic use. In other words, the first economic application of an invention is called innovation. Clearly, the point of distinction is that the former has no economic significance until it is used to produce goods or obtain a market value. This view of the problem can lead to inappropriate dissociation of the terms. In the transcendentalist approach, for instance, the emergence of an invention is attributed to occasional inspiration and personal effort, which does not justify firms' expenditure on R&D. Usher

(1954) criticized this ahistorical view and pointed out the possibility of resource allocation to research as a way to influence the process of technical change, which must be interpreted in the light of firms' competitive environment. In competitive markets, the probability of an invention occurring is intentionally enhanced in order to generate innovations. Thus, invention is a process in which resources are deliberately allocated to research to bring about the possibility of innovation. The discussion of this matter will naturally lead to the theoretical debate on the origin of innovations. This is a broad area in the literature of technical change. After the Second World War, theoretical advances have been associated with two basic lines of discussion: the 'demand pull/technology push debate' and 'induced innovation'. The basic ideas of these approaches are presented here.

In traditional neoclassical thought, technology was regarded as an exogenous phenomenon and the generation of innovations was dissociated from economic factors. Few economists during the eighteenth and nineteenth centuries attempted to make a link. Marx (1867), who incorporated technical change as an endogenous element of the capitalist society, made a significant contribution, but he did so in a way paradigmatically apart from the prevailing literature. The first major dissension from the neoclassical school can be attributed to Schumpeter's (1934, 1943) works on long-term economic development and structural change in capitalist societies.

In Schumpeter's early work, the entrepreneur has a central place in the economic system. Entrepreneurs are always searching for a monopolistic position in the market in order to earn extra profits. The most innovative ones are able to create a temporary monopoly by introducing new ideas into economic life. However, as soon as competitors start to imitate their innovation, profits are eroded by increasing competition. The only way to maintain a monopolistic position is by generating a constant flow of new and disruptive ideas. Although this view of economic development was certainly a departure from the neoclassical idea of equilibrium, the notion that perfect competition leads to a social optimum was still present. In his later work, however, Schumpeter changed his view of the economic system and highlighted the importance of monopolies to the capitalist economy. He argued that the generation of innovations requires massive investment of resources in R & D, large minimum size and sometimes positive returns of scale, which only monopolies and oligopolies are in a position to support. The small competitor of the neoclassical perfect market would be unable to generate the flow of new ideas that characterised the development of modern capitalism.

The Schumpeterian theory established a direct link between technology and economic growth.

> Technology, whether generated outside the economic system or in the large R & D laboratories of a monopolistic competitor is for Schumpeter the leading engine of growth. Therefore the technology push hypothesis of the origin of innovations finds a natural place in Schumpeter's ideas (Coombs et al., 1987, p.95).

Schmookler (1966), arguing in favour of demand forces, investigated time series data on patents and investment in capital goods and found the latter tending to lead the former. Moreover, fluctuations in investment were better explained by external events than by the flow of inventions. He concluded that upward innovation flows were explained by increasing investment. These findings formed the basis for the 'demand pull' hypothesis about the origin of innovations. Most theoretical advances achieved in the technical change area were greatly influenced by the 'demand pull/technology push' debate that follows Schumpeter and Schmookler's ideas.

The second analytical line in the literature refers to 'induced innovations'. The original formulation of the induced innovation theory is attributed to Hicks (1932, p.124), who stated:

> A change in relative prices of the factors of production is itself a spur to invention, and to invention of particular kind - directed at economizing the use of a factor which has become relatively expensive.

Hicks was interested in income distribution in a macroeconomic perspective. Later, the theory shifted towards the microeconomic approach that characterized the induced innovation theory. Salter (1966) argued that at competitive equilibrium all factors are paid according to their marginal value; therefore, all factors are equally expensive to firms. The entrepreneur is interested in reducing costs, no matter which particular factor has to be saved. A change in a factor's price leads to substitution, not technical change, which re-establishes the equilibrium and, consequently, one factor is no more expensive than another. 'When labor costs rise, any advance that reduces total cost is welcome, and whether this is achieved by saving labor or capital is irrelevant' (Salter, 1966, p.43).

Salter's induced innovation was favourably received by economists (Thirtle and Ruttan, 1987) and other approaches to the theory were subsequently developed. Kennedy (1964), von Weizsacker (1966) and Drandakis and Phelps (1966) found that there was a greater tendency to save a particular factor the greater its share of total costs. The factor's share, not its price, determines the inducement to technical change. In both cases, anyway, 'inventive and innovative activities have become at least partly endogenous to the economic system' (Coombs et al., 1987, p.105). They failed, however, to include the costs of inventive and innovative activities in their analysis. Further developments attempted to add to the basic model the relative cost of obtaining different types of technology. On this matter Thirtle and Ruttan (1987, p.32) summarized the following results:

> 1) Any rise in the expected present value of the total cost of a factor will lead to an increased allocation of resources to the research activity that most saves that factor.
> 2) A rise in the cost of research that saves a particular factor or a decline in the productivity of that research will reduce the allocation to that line of research, and hence bias technical change in the direction of the other factor.
> 3) With no budget constraint on research activities, a rise in the value of output (due to greater output or higher price) will increase the research budget and hence the rate of productivity growth.

This version, which portrays a broader induced innovation theory, does more justice to this theoretical line than Hicks' first attempt. A different approach was suggested by Rosemberg (1969). Instead of focusing on market-price-based reasoning, he stressed the importance of other inducements, such as bottlenecks in technological development, scarcity of raw materials and political factors. Hayami and Ruttan (1973), however, argued that bottlenecks would be reflected in relative factor scarcities which are signalled by market prices.

The evolutionary theory of technical change

In the evolutionary approach, technical change is seen as a disequilibrium process. In this sense, it differs from the induced innovation theories and is closer to Schumpeter's ideas. In general, followers of disequilibrium models criticize the Neoclassical School for its equilibrium premise. They argue it is an ad hoc and false assumption taken from physics and not pertinent to economics. The Neoclassical School's response is that the evolutionary approach borrowed its concepts from Biology, so the same criticism would be valid. Nelson (1987, p.12) defends the evolutionary framework arguing:

> Most emphatically we do not mean blindly picking up ideas and models from biology. While we would agree with Marshall that there is much that

> economists can learn by looking to what the biologists do, social, economic and technical change must be understood on their own terms. Thus, by an evolutionary theory we mean to include a relatively large class of models of change, with evolutionary theory in biology being a special case, and evolutionary theory of technical change being another special case.

Nelson and Winter (1982) provided an influential synthesis of this line of thought. The evolutionary theory contains two fundamental mechanisms that are incorporated into economic theory. The first mechanism generates 'novelties'. In contrast to biology, in economics variety is generated through firms' deliberate actions as a consequence of competition. How this generation process works is what the evolutionary theory tries to describe. The environment in which firms operate plays an important role in determining their behaviour when they compete with other firms. So institutional and cultural elements of the system, as well as historical evolution, are fundamental to understanding the process of change. In biology, survival of a specie depends on its capacity to fit into the environment. Organisms develop natural and hereditary features that enable better adaption. There is a natural 'mechanism of selection', in which the less able is threatened by extinction. In economics, firms in competition develop new products and new methods as a way of surviving within the market. The more successful they are, the bigger will be their importance within the economy. However, unlike biological natural processes, firms are able to create competitive capacity through both deliberate search and market expectations. In such a way, firms incorporate routines that condition their behaviour when the environment changes. Technical routines are developed for different purposes - for instance, investment policy and research activities - and form the operational conditions of evolutionary models.[1]

Nelson and Winter (1982) modelled economic growth on an evolutionary theory of innovation. They built a complex computer simulation model that presented similar results to Solow's (1957) neoclassical framework. However, as they emphasised, an evolutionaty formulation

> must provide an analysis that at least comes close to matching the power of neoclassical theory to predict and illuminate the macroeconomic patterns of growth. And it must provide a significantly stronger vehicle for analysis of the process involved in technical change, and in particular enable a fruitful integration of understanding of what goes on at micro level with what goes on at a more aggregate level (Nelson and Winter, 1982, p.206).

The outcome is a fuller understanding of the economy and its evolution.

Coombs et al. (1987), referring to Nelson and Winter's (1982) model, pointed out that

> their work represents a general tendency in the literature to combine microeconomic inducement mechanisms, managerial models of firm behaviour, the structured fields of technological possibilities which are common to large number of firms but which are not sufficiently powerful to generate identical responses from firms due to the complexity of selection enviroments, and uncertainty (Coombs et al., 1987, p.120).

Smith (1991) simulated evolutionary growth in a multisectoral economy. In his model, firms' initial behaviour was supposed to be governed by three groups of variables. The first group, representing the external environment, comprised variables relating to market prices, wage rate, interest rate, industry demand and market share. The second, described the economic and physical state of the firm, such as capacity and technology, stocks, firm's prices, planned new capacity, funds and debt, and search environment. Finally, the third group consisted of variables that determined the firm's decision rules (routines): utilization rate, desired utilization rate, desired investment rate, and price markup. Once the initial conditions are defined, firms decide about production and investment. Investment funds can be spent either on search or on new capacity. As a result of those decisions, a feedback mechanism comes into operation and the firms' rules are subsequently assessed. A time trend for the variables can be built to allow for assessment of firm, industry and whole economy performance. The model also allows an investigation of performance under different scenarios - for instance, assuming selection either with or without imitation and search. An interesting result was that the introduction of imitation increased the speed of diffusion and generated a sigmoid curve such as observed in empirical studies, as will be seen in this chapter. The role of different external environments could also be evaluated by simulation procedures.

Innovation, adoption, and diffusion

The evolutionary model gave a hint of the difficulty of separating analytically the origin of innovations and their diffusion. In the evolution of a technology, the maturation process implies diffusion, which constrains subsequent generation of innovations. However, many studies are specifically dedicated to the analysis of adoption and diffusion of innovations, taking the origin of technology as given. We shall review some of these in the next three sections. Before proceeding, it will be useful to discuss some aspects of the adoption and diffusion concepts.

A distinction must be made between adoption and diffusion. Adoption studies try to explain why firms adopt an innovation in a particular moment in time. The focus is the firm and its behaviour. Conversely, most diffusion studies are devoted to explaining the aggregate adoption over time, without considering the foundations of the single firm's decision process. In this view, adoption is a static version of the problem while diffusion deals with the dynamic. However, the distinction between adoption and diffusion is more difficult to make. Adoption can explain why some firms adopt early, while others delay doing so. In that case, a diffusion curve can be obtained, by aggregation, from a microeconomic modelling of the adoption decision. Studies of this nature are quoted by Feder et al. (1985) as aggregate adoption models. The threshold models, as will be seen, are examples of how the diffusion process can be obtained from individual adoption analysis. Chapter 7 reviews recent empirical studies that used duration analysis technique to offer this dynamic perspective. In fact, studies that explain adoption over time also explain the diffusion process.

Another distinction that appears in the literature refers to 'inter-firm' and 'intra-firm' diffusion. Once a firm has adopted an innovation, it may be useful to know to what degree the new technology has displaced the old one. For example, instead of a massive replacement, firms can progressively substitute the old capital. In the early stages, the innovation would be only partially adopted. This phenomenon is used to differentiate between inter-firm diffusion and intra-firm diffusion. The former is related to 'the rate of imitation - the rate at which firms begin to use an innovation' - while the latter is concerned with 'the rate at which a particular firm, once it has begun to use a new technique, proceeds to substitute it for older methods' (Mansfield, 1968, p.173). In adoption of agricultural innovation studies, for instance, this distinction is useful to differentiate farmers that have fully adopted the Green Revolution's technological packages from farmers that have incompletely adopted them.

The epidemic model

Innovations are generally assumed to be advantageous in comparison with the existing technology. However, it is rare for them to be adopted immediately by all potential adopters. Early studies pointed out lack of information, always related to risk, as the main reason for some firms delaying adoption. Once the first adoption has occurred and as time passes, the number of adopters increases and, consequently, information on the use of the new technology is continuously accumulated. As the risk associated with adoption decreases during this process, the number of firms desiring to adopt is enlarged. However, since the number of potential adopters is limited, two opposite forces rule. First, the number of adopters acts favourably on diffusion. However, as the proportion of adopters increases, the number of potential adopters falls, creating the second and counteracting force. As a result, a bell-shaped frequency distribution for the number of new adoptions over time is obtained. This process, known as the epidemic model, is similar to the spread of an infectious disease. Uninfected persons catch an illness by contact with the population already infected. The epidemic proceeds, but opposing forces are in play. As the number of infected persons rises, the probability of new contacts increases. However, as the population is fixed, the number of uninfected persons decreases. Analogously, in the diffusion of an innovation, contacts allow for the spread of information and, consequently, a diffusion process is obtained.

The model can be represented by the following equation:

$$\frac{dm_t}{dt} = \beta(n - m_t)\frac{m_t}{n} \qquad \beta > 0 \tag{5.1}$$

Where m_t is the number of individuals having adopted at the time t and n is a fixed population. The number of individuals adopting an innovation in the shortest interval of time after t depends on the size of the uninfected population (potential adopters), the population already infected (adopters at time t), and β. This latter parameter depends on the attractiveness of the innovation and the efficiency of the communication channels, which are assumed as given during the diffusion process; thus, β remains constant.

The number of adopters over time can be represented by a bell-shaped frequency distribution (Figure 5.1). Integrating equation 5.1, we have

$$m_t = \frac{n}{1 + \exp(-\alpha - \beta t)}, \tag{5.2}$$

where α is the constant of integration. This is the cumulative density function of a logistic frequency distribution, which can be represented by a symmetric sigmoid (S-shaped) logistic curve (Figure 5.2). The parameter β depicts the speed of diffusion. The larger it is, the faster is the diffusion process. The constant of integration α defines the position of the curve on the time axis. Estimates of parameters are easily obtained from the following transformation of equation 5.2:

$$\log(\frac{m_t}{n - m_t}) = \alpha + \beta t \qquad 5.3$$

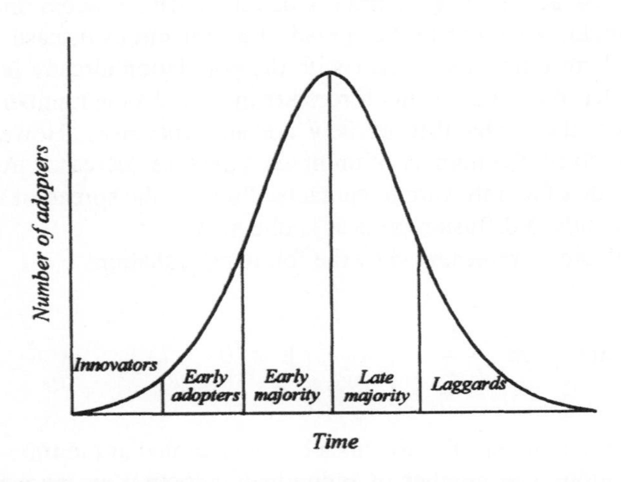

Figure 5.1 Logistic frequency distribution and adopter categorization

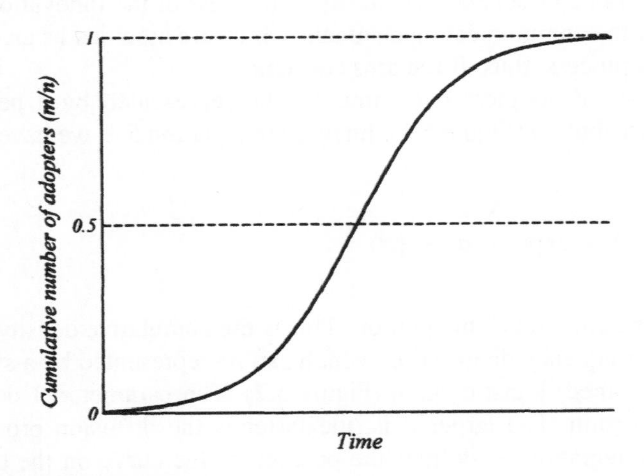

Figure 5.2 Sigmoid logistic curve

Rogers and Shoemaker's categories of adopters

Rogers and Shoemaker (1971) divided the adoption distribution curve into five adopter categories (Figure 5.1). Initially, a few 'innovators' start to use the new technology. They are venturesome and have a strong desire to try new ideas. They are followed by a second category, 'early adopters', who exhibit a successful and discrete use of innovations. They are respected by their counterparts and represent a model to follow. The third category is the 'early majority', which comprises individuals who think carefully before adopting a new idea and rarely assume a leadership position. The following category is formed by a 'late majority' of individuals, who are sceptical about innovations and wait until they have been widely diffused. Finally, 'laggards', the last to adopt an innovation, have traditional values and tend to look to the past as a point of reference for their decisions. When they adopt an idea, it is no longer new and innovators are probably already trying another novelty. Rogers and Shoemaker (1971), after analysis of several diffusion studies, characterize earlier adopters in relation to later ones:

> The relatively earlier adopters in a social system tend to have more education, a higher social status, more upward social mobility, larger units, a commercial rather than a subsistence orientation, a favorable attitude toward credit, and more specialized operations. Earlier adopters also have greater empathy, less dogmatism, greater ability to deal with abstractions, greater rationality, and more favorable attitudes toward change, risk, education, and science. They are less fatalistic and have higher achievement motivation scores and higher aspirations for their children. Earlier adopters have more social participation, are more highly integrated with the system, are more cosmopolite, have more change agent contact, have more exposure to both mass media and interpersonal channels, seek information more, have higher knowledge of innovations, and have more opinion leadership. They usually belong to systems with modern norms and to well integrated systems. (Rogers and Shoemaker, 1971, p.195)

Further developments of the epidemic model

The simple epidemic model (equations 5.1 and 5.2) suffers from two basic limitations. First, there is only one explanatory variable, cumulative adoption, which is used as a proxy for a word-of-mouth internal mechanism of information diffusion. Neither external information, such as suppliers' selling strategies, nor other variables related to the demand side enter directly in the model. Secondly, it relies on aggregate data, instead of data from individual firms, which would better explain the microfoundations of the decision process. This matter was considered in later developments of the theory.

Griliches' (1957) study on the diffusion of hybrid corn in USA was the first well-known work in economics to use the epidemic model. Instead of using only the cumulative adoption approach, he applied a two-step econometric technique to explain β. With this procedure, other explanatory variables could be introduced in the model. In the first step, the logistic curve was fitted for 31 American States by using ordinary least squares. Different diffusion paths were generated (one for each State), resulting in 31 values for α and β. In the second step, the variance of β was explained through a linear regression on three independent variables: corn acres per farm, pre-hybrid yield, and the difference in yields between the old (open pollinating seed) and the new (hybrid varieties) technologies. These variables were assumed to be a proxy for the profits obtained through technological change. They formed the demand side of the problem. The supply side was portrayed by the parameter α, which set the date of the origin (year at which 10% hybrid corn was planted). Lags were explained by differences in profitability to seed producers, which decreased with the distance from the Corn Belt.

Mansfield (1961) used the same approach to analyse the diffusion of twelve innovations in USA manufacturing industry. He found the speed of diffusion was positively related to the profitability of the new technology, but inversely related to the necessary investment. In subsequent work, Mansfield (1968) tested four additional factors affecting the diffusion speed. First, if the new technology is set to replace durable equipment which has not yet finished its planned payoff period, the firm will be reluctant to discard it. In other words, the younger the old technology, the lower the 'rate of imitation' (diffusion speed). Second, where the rate of output growth is high and firms are expanding, then new plants will be built incorporating the innovation. Here, a positive relationship is expected between increasing output and diffusion speed. Third, the parameter β can increase over time as a result of improvements in the communication channels, better methods of evaluating machinery replacement, or because there is a move towards accepting frequent changes in techniques. Finally, one must consider the phase of the business cycle in which the innovation is introduced. Additional explanatory factors for the diffusion speed were tested in other studies. The most common were variables representing 'industry structure' (Lissoni and Metcalf, 1993).

The mathematical form assumed for the sigmoid curve was also a subject of criticism. Most empirical studies confirmed an S-shaped (sigmoid) curve as the best representation of the diffusion time-path. However, the logistic curve is not always the closest approximation to reality. Dixon (1980) argued the inapplicability of the logistic curve in cases where data exhibit a significant degree of asymmetry. He studied the hybrid corn diffusion in the USA and found the Gompertz[2] curve fitted better than the logistic curve for some States. Moreover, Stoneman (1983) stressed that logistic and Gompertz curves are not the whole class of curves which are able to represent a S-shaped one. Since the cumulative S-shaped curve is derived from a bell-shaped frequency distribution, other S-shaped cumulative density curves can also be derived from alternative bell-shaped distributions. In the lognormal distribution, for example, the inflexion point is a function of the variance and the mean distribution.[3] As a result, a set of sigmoid curves can be derived, each with different inflexion points occurring at any value of n_t /m.

Although Griliches and Mansfield had produced sensible economical explanations by adding new variables to the epidemic model, and a variety of S-shaped curves had been suggested, two problems remained. First, the data used in the models referred to the industry average values. Thus, firms (adopters and potential adopters) were assumed to be homogeneous at the industry level. Heterogeneity among individuals could not be controlled and the models still lacked a more appropriate microeconomic empirical analysis. Second, given data constraints, the number of explanatory variables in the second step of estimation is usually constrained by the limitation imposed by problems with degrees of freedom.

Threshold models

In David's (1966) study on innovation diffusion, the individual firms' adoption decision was at the centre of the analysis. He explained why, for the mechanical reaper, so many years elapsed between McCormick's patent in 1834 and the widespread adoption of the reaper in the 1950s. During the first half of the nineteenth century the amount of grain that a farmer could cultivate was limited by harvest labour availability. In other words, the labour intensive method of harvest combined with labour scarcity imposed a constraint on output production. The mechanical reaper, as labour saving embodied technology, could solve the problem. Although it was available, immediate widespread adoption did not occur. David argued there was no constraint on the machinery supply side, the reason for delay being insufficient farm acreage. Given the indivisibility of the machine and the absence of cooperative arrangements, only large farms, in which cost per acre could fall, had initially adopted. Since most of the farmers at the time were small, the labour saving resulting from the innovation did not repay the initial investment required to buy the mechanical reaper.

Davies (1979) reproduced David's framework as follows. Reaper adoption will be profitable if

$$w_t(L_{iOt} - L_{iNt}) \geq p_{iNt}\,, \tag{5.4}$$

where w_t is the wage rate, L_{iOt} and L_{iNt} are the annual labour requirements of the old and the new methods in the farm i at time t, and p_{iNt} is the average annual cost of a mechanical reaper. If there are no scale economies,

$$\begin{aligned} L_{iOt} &= a_1 S_{it} \\ L_{iNt} &= a_2 S_{it} \qquad a_1 > a_2 > 0 \end{aligned} \tag{5.5}$$

where a_1 and a_2 are technical coefficients and S_{it} is the size of the farm i at time t. Substituting 5.5 into 5.4 gives the condition for profitable adoption:

$$S_{it} \geq \frac{p_{iNt}}{w_t} \frac{1}{a_1 - a_2} \tag{5.6}$$

Instantaneous diffusion of the mechanical reaper did not happen because the expression 5.6 did not hold for all farms at the time it became available. As can be seen from the equation, diffusion would depend on farm size, the innovation cost relative to wage, and technological improvements to the reapers. Adoption of the new technology was unprofitable in comparison with the old method below an certain acreage. However, this ‘threshold’ level would change over time as a consequence of changes in relative prices and technical parameters.

Davies, in fact, had re-established David's model using the Probit Analysis, which, at that time, was being used in the study of the diffusion of durable goods (Cramer, 1969, and Bonus, 1973). According to these studies an individual consumer will be found to own goods at any time t if his income y_{it} is greater or equal to a threshold income level y^*_{it}. Ownership can be represented by a dummy variable as follows:

$$q_{it} = 1 \quad if \quad y^*_{it} \leq y_{it},$$
$$q_{it} = 0 \quad otherwise.$$

The interest here is in obtaining a relation between the probability of ownership $P(y^*_{it} \leq y_{it})$ and income. Both individual income and individual threshold income can change over time. For instance, consumers who have not bought a good can reverse their decision because either their income has increased or their threshold level (preferences) has decreased.

Davies' approach to the adoption decision followed behavioural assumptions regarding individual potential adopters. The decision to adopt is taken when a related set of variables reaches a certain level. In other words, there is a 'threshold' level to cross. In equation 5.6, for instance, the adoption decision depends on farm size, relative prices and technical coefficients. Figure 5.3 shows the relative frequency distribution of farm size and the threshold farm size S^*_{it} at a moment in time. As time passes, the mean farm size increases, shifting the distribution curve to the right. As a result, the number of adopters is increased and a sigmoid diffusion curve is generated. Alternatively, or simultaneously, an increase in wages relative to the cost of the reaper shifts S^*_{it} to the left, allowing also for the diffusion process. Here, we have a clear difference between epidemic and probit models. In the former, the diffusion curve results from time, as a proxy variable, while in the latter it is explained by microeconomic characteristics.

Davies' (1979) own model is similar to the above approach. He introduced post innovation aspects, such as suppliers learning and enhanced quality information, in the basic framework and considered the role played by risk. Following his notation, the adoption decision is taken by firm i according to both an expected payoff period associated with the innovation at time t, ER_{it}, and its maximum acceptable payoff period, R^*_{it}.[4] Let

$$q_{it} = 1 \quad if \quad ER_{it} \leq R^*_{it}$$
$$q_{it} = 0 \quad otherwise,$$

where $q_{it}=1$ is an adoption decision and $q_{it}=0$ is non-adoption.

ER_{it} was specified as a function of firm size and other characteristics, such as technical attributes and ability to collect and process information related to innovation. It decreases monotonically over time for cheap and technologically simple innovations. The anticipated profitability of adoption increases due to improvements in technology through 'learning by doing' on the suppliers' side.[5] Also, information from firms which have already adopted, from suppliers, and from 'active search' is supposedly enhanced with the passage of time. The expected payoff period and risk are reduced by both post innovation technical improvements and better quality information.

Similarly, the threshold payoff period, R^*_{it}, was set as a function of firm size and variables related to risk, such as the age of farm management, degree of internal financing, profit trends, growth performance, etc. For cheap and simple technologies, R^*_{it} increases monotonically over time due to the reduction on the 'risk premium' and the 'effect of competitive pressures'.

Those characteristics of less expensive and simple technologies result in a positively skewed lognormal diffusion curve, while for more expensive and technologically complex innovations, to which those attributes do not apply, a symmetric cumulative normal curve is obtained. Thus, David provided an explanation for alternative possibilities of S-shaped curves. Technological characteristics, firm characteristics, and industry characteristics are important in determining the sigmoid form. Also, the supply side can be influential. The firm selling the technology, or governmental institutions, can accelerate the rate of diffusion in the early stages through marketing mechanisms, such as advertising, salesmen, and extension agents.

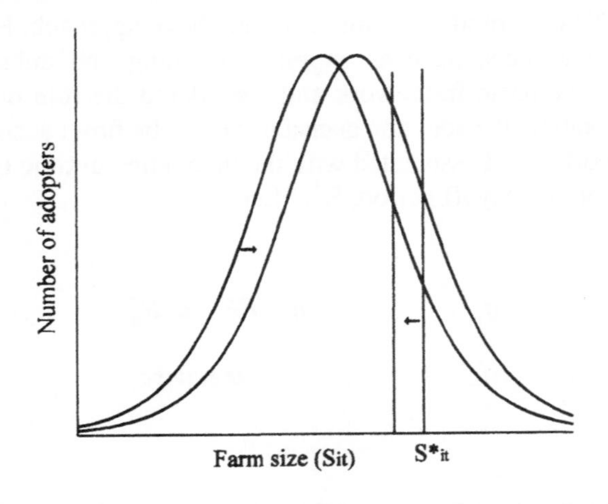

Figure 5.3 Frequency distribution of farm size

The Bayesian model

Stoneman (1980) proposed a new approach to Davies' interfirm diffusion model. As emphasised by him,

> The advantage of the model over other probit models is that the critical value of the characteristic determining the use of new technology is determined by the rational maximization behaviour of individual entrepreneurs rather than being generated in an ad hoc manner (Stoneman, 1980, p.179).

According to the model, an individual entrepreneur has two available technologies: the new and the old. Both techologies have perceived returns, which are reflected in subjective probability distributions. The author showed that the greater the initial estimation of the mean return of the innovation, the earlier will the firm be expected to adopt. On the other hand, the higher the returns from the old technology, the later will the new technology be introduced. As entrepreneurs differ in their evaluation of returns and in their attitudes to risk, there will be different time to adoption. Consequently, the probability that a firm will be using the innovation at any time can be given by a distribution function. Thus, a diffusion path can be generated and, if this distribution has a bell-shaped form, the diffusion curve will be sigmoid.

The game theoretic approach

Reinganum (1981a and 1981b) has built an approach in which diffusion is explained purely by firms' strategic behaviour. 'Even with perfect information and identical firms, there will be a diffusion of the new technology rather than simultaneous adoption' (Reinganum, 1981b, p.618). It is supposed that firms can make more profit using the new technology than the old, and that the first adopter is expected to make substantial profits at the expense of the others. However, due to declining adjustment costs, later adopters may save money on the cost of purchasing the new technology. Thus it is not advantageous for identical firms to bring the new technology on line at the same time. The firm must weigh the costs and benefits of delaying adoption, and take into account its rivals' strategic behaviour. The author used game theory to formalize the problem in a model for an industry of two identical firms which manufacture a homogeneous product. He established the interesting result that:

> ... one firm will adopt the innovation at a relatively early date, the other relatively later. Thus, even in the case of identical firms and complete certainty, there is a diffusion over time (Reinganum, 1981a, p.395).

The impact of market structure upon the diffusion process was investigated by increasing the number of firms in the model. It was shown that the difference in profit rates immediately preceding and immediately following adoption are important for explaining the adoption date schedule.

Adoption of agricultural technologies

In the threshold, Bayesian, and game theoretic models presented above, firms are assumed to behave optimally. These are equilibrium models of diffusion in which, at any time, a firm will decide to adopt an innovation if this decision generates profit. Most, if not all, empirical studies on agricultural technology adoption have been dominated by equilibrium models. Many of these studies, inspired by Griliches' (1957) and Roger and Shoemaker's (1971) models, have tried to explain the S-shaped pattern of diffusion, while a few of them have attempted to build a Bayesian approach.[6]

Feder and O'Mara (1982) formulated an aggregate innovation diffusion model based on the assumption that individual farmers revise their beliefs in a Bayesian fashion. They showed that Bayesian learning can generate S-shaped adoption function for a dominant innovation. Lindner et al. (1979) used a similar approach to demonstrate that the more profitable the innovation, the shorter the time to adoption. Their model showed that initial expectations about the relative profitability of an innovation affect the time to adoption, which lengthens according to the decision maker's initial degree of pessimism.

A common approach has been to examine of the effect on adoption of several economic and non-economic variables. The appropriate way to test hypotheses about these variables would be through the analysis of individual farmers' decisions over time, such as in the threshold model. However, most studies deal with cross-section data and have used static econometric procedures. A dichotomous adoption variable (adoption, non-adoption) is explained by a set of explanatory covariates. The most commonly used econometric methods are correlation analysis, ordinary least squares, and logit/probit models. It will be shown in chapter 7 that the usual logit /probit models are inadequate to explain adoption from a dynamic perspective. They explain adoption as a static process, that is, at a given point in time, and do not take into account either the adoption date or the time path of explanatory variables. As will be seen, duration analysis has recently been used to perform a more appropriate examination of the individual decision process over time.

Most Green Revolution packages have divisible components such as fertilizers and pesticides. Farmers may apply a full prescription in all fields or use a small quantity on a few crops. Where it is important to examine of the extent and intensity of adoption, a dichotomous dependent variable may not be the best approach and the Tobit model should be used instead (Akinola and Young, 1985).[7]

Several determinants of adoption have been studied in the agricultural economics and rural sociology literature. According to Feder et al. (1985), the most common explanatory variables are farm size, risk and uncertainty, human capital, labour availability, credit constraints, tenure, and supply constraints. For example, Hill and Kau (1973), modelling grain dryer purchasing decisions, hypothesized farm size, farm type, method of harvest, farm ownership, shelled corn storage capacity, percentage of grain sold at harvest, and age of the farmer as the most important variables determining adoption.

The role played by farm size depends on technical, economic, and institutional factors. Some technologies require large fixed costs for implementation, which are inaccessible to small farmers. Examples of these barriers are expensive indivisible capital embodied technologies (such as harvesters) and the high cost of learning and training hired labour. Evidence suggests that larger farms adopt early, while smaller ones lag behind. One reason for this is credit constraints. However, small farmers can adopt early if the available acreage induces them to farm more intensively in order to survive; or, alternatively, if they are able to work more efficiently using low cost family labour. In addition to this factors there are others: risk, access to inputs, wealth, access to information, and so on, which are related to farm size and so form part of the environment that determines adoption behaviour. For instance, a constraint on credit can play an important role in the adoption decision although, in particular cases, lack of external financial resources can be overcome by off-farm income.

The adoption decision is largely influenced by the uncertainty that arises from novelty. Elements such as susceptibility to pests, climate adaptability and input availability are matters of subjective assessment, unless full information is available. Risk is a crucial explanatory variable when a decision must be taken concerning the introduction of a technique for which yield is not yet known. Measurement is difficult and, consequently, empirical studies use proxies for doing so. Dummy variables are applied to represent climate, topography, fertility, disaster (infestation, droughts, floods), and markets. All of them are supposedly related to uncertainty and are used as a proxy for risk. Also, the more exposed to information farmers are and the more able they are to understand it, the less the subjective uncertainty. Variables representing extension services are the most commonly used in this case. There are other proxies, such as exposure to mass media, literacy, level of education, time spent out of the village, and observation of adopters. Bayesian models consider farmer experimentation or 'learning by doing' an important source of information.

According to the traditional models of diffusion, once the information is available, farmers will make a choice that reduces time and physical labour, and, simultaneously, increases productivity and efficiency. However, if access to sufficient capital, land, credit, or other economic resources is difficult, adoption can be constrained, even if adequate access to information is available. Economic barriers may prevent adoption and factors related to information may play a secondary role.

Another factor explaining adoption is human capital. Generally, farmers' abilities (work ability and allocative ability) are supposed to play an important role in determining returns from agriculture. Following this line, Rahn and Huffman (1984) suggest that farmers' investments in schooling, experience, information, and health, are expected to enhance allocative skills and increase the efficiency of adoption decisions. Also, age, assumed as a proxy for farmer experience, is supposed to be positively related to the likelihood of adoption.

Technologies can be labour saving or labour using. They can also increase the seasonality of labour demand. Consequently, availability of family labour or feasible supply of hired labour during peak seasons, is taken as another factor in determining adoption. Moreover, not only labour availability, but also ease of access to other complementary inputs, such as seed, fertilizers, water, and storage capacity, are sometimes relevant. Finally, a distinction must be made between tenants and owners. Generally, the latter are expected to be more receptive to innovation. However, empirical evidences are conflicting, as tenants' interest in innovations may depend on the possibility of access to credit, input markets, product markets, and technical information. In conclusion, factors that affect adoption cannot be analysed independently, but must be related to the institutional and economic context in which the innovation is to be introduced.

Conclusion

Debate on the origins of innovation has led to a broader theory of technical change and allowed for the integration of the macro and micro perspectives of the economic system. The latter is the central focus of the evolutionary school, which combines the Schumpeterian approach with the behavioural theory of the firm.

The adoption and diffusion literature is evidence of continuous progress since the epidemic model was first applied in the 1950s. Recent studies have been more concerned with individual decision behaviour modelling rather than with exclusive examination of aggregate diffusion. Advances brought about by the Threshold, the Bayesian, and the Game Theoretic approaches permit an explanation of aggregate diffusion through the micro elements of the adoption decision. Most empirical studies on adoption, however, have applied statistical methodologies which are, in essence, static. Only recently, as will be seen in chapter 7, has a dynamic statistical approach (duration analysis) been used to test the adoption decision over time.

The literature on agricultural technologies shows that there are many factors affecting adoption of Green Revolution innovations. The effect of different variables on the adoption decision depends on the general context in which the new technology is introduced. Some economic variables, such as price and size, are usually important, but sometimes their effect can be influenced by institutional arrangements, such as credit constraints and tenancy.

Notes

1 For a discussion of the key elements of the evolutionary approach see also Saviotti and Metcalfe (1991).

2 The Gompertz curve has an inflection point at $m_t/n = 0.37$, assuming a long-tailed diffusion path. In other words the point of inflexion occurs prior to the date at which 50% (logistic case) of the firms have adopted the innovation.

3 In the lognormal, the mode occurs at $x = e^{\mu-\sigma^2}$, where μ is the mean and σ^2 is the variance of the distribution.

4 ER is taken as a inverse measure of the expected profitability of adoption.

5 Learning by doing can be interpreted as a process of enhancing knowledge through experience, which is important in reducing uncertainty and risk. For an accurate definition and theoretical considerations see Arrow (1962).

6 For a review of empirical and theoretical studies concerning adoption of new technologies in agriculture see Feder et al. (1985) and Thirtle and Ruttan (1987, section 3.4.2).

7 See Maddala (1983) and Greene (1993) for a mathematical understanding of the Tobit model, and Akinola and Young (1985) for an empirical analysis using this methodology.

6 The determinants of the adoption of sustainable agricultural technologies

Introduction

Adoption of sustainable agricultural technologies has only recently become an area of study in rural sociology and economics. Many sustainable practices are in the early stage of diffusion and analyses of their adoption are not abundant. The literature, however, has continuously increased in recent years. Some empirical studies have investigated the influence of different factors on the adoption decision, while others have been devoted to the analysis of possible barriers to wide diffusion. Analyses of the economic effectiveness of different production systems have been extremely important in evaluating the economic appeal of sustainable approaches. Other studies, such as those focusing on farmers' environmental attitudes, have examined non-economic factors. The purpose of this chapter is to review the literature concerned with the determinants of the adoption of agricultural sustainable technologies.

Adoption of sustainable technologies in agriculture: economic vs non-economic reasons

Studies of farmers' environmental attitudes have been conducted to test the hypothesis that urban residents are more environmentally concerned than their rural counterparts. Two basic explanations are generally given. First, urban residents have greater concern because they are often exposed to a higher level of environmental degradation. Second, rural residents are supposed to be less environmentally concerned because they have predominantly extractive occupations and are more dependent on the use of the natural environment.[1]

During the 1970s, empirical studies revealed two different results on this matter. Those focusing on problems at the local or community level generally presented stronger confirmation that rural people are less concerned with the environment than urban dwellers. In contrast, when state or national environmental problems were considered, there was weak support for that hypothesis (van Liere and Dunlap, 1980; Tremblay and Dunlap, 1978).

Although evidence from the 1970s suggests farmers are less concerned with the environment, it cannot be generalized. Many farmers are adopting environment-friendly technologies. Why are they doing this? Do they, or their farms, have special characteristics? Are they really more concerned about the environment or rather are they pursuing profits and taking environmental measures to get them?

The first studies dealing with adoption of sustainable technologies in agriculture date back to the 1970s. Wernich and Lockeretz (1977) used a sample of organic farmers to address the basic question set by most empirical studies on technology adoption: why do agents choose a specific production technique? The methodological approach used in the analysis was to estimate the frequency distribution of farmers' attitudes and practices. The survey showed most individuals were conventional farmers and, after switching to an organic system, they continued to farm on a commercial scale. Ideological or philosophical considerations played a secondary role. The main reasons for the adoption of organic methods were problems with conventional practices, such as livestock and human health, soil deficiency, cost, and non-effectiveness of chemicals. The authors concluded that those farmers did not differ, in a 'fundamental way', from conventional ones. This conclusion challenged the popular image that organic farmers used animal power, grew fruits, vegetables and speciality crops for the organic or natural food market, and operated subsistence enterprises. Here, an important contribution was given to challenge the prevailing view. The study, however, does not reveal whether organic farmers could be considered profit maximizers, which is a general assumption in most studies of conventional farmers' adoption behaviour.

Profit maximization is a standard assumption in neoclassical production economics. In the analysis of technical change, the starting point is the production function. The farmer, who is led by the profit maximization rationale, must choose between alternatives to combine inputs to generate outputs. Other values such as independence, desire to protect the land, risk aversion, etc. are not included in the standard models and are assumed as external fixed variables.

Daberkov and Reichelderfer (1988) used the neoclassical approach to obtain some insights on the prospects for low-input agriculture in the USA. Data covering the postwar period up to the 1970s' energy crises show that fertilizer and pesticide prices decreased relatively to the wage rates, farm machinery prices, and land cost. As a result, those inputs became an attractive substitute for labour, capital and land. In addition, commodity programmes induced farmers to use external inputs more intensively. Agrichemical usage expanded in response to farmers' rational profit-maximizing behaviour. Under these conditions, low-input systems, which imply substitution of land, labour, management, and information for agrichemicals, had no chance of being adopted, unless relative prices and/or commodity programmes changed. If low-input farming practices were adopted, one can conclude their adopters were not profit-maximizers. Daberkov and Reichelderfer (1988), however, suggested that under specific conditions (which were ignored by their general model) adoption could be explained. In fact, low-input farmers were profit-maximisers and opted for these farming systems because of microlevel conditions, such as scale, output level, credit constraints, management ability, education, information, etc. Although there was little market incentive for adoption of low-input agriculture systems, some farmers adopted it because, under particular conditions, it was economically beneficial to do so.

Oelhaf (1978) took a different view on this matter. According to him, there were favourable economic conditions to produce organic food in the USA during the 1970's, but for most organic farmers the commercial incentive played a secondary role.[2] Some of them consciously rejected modern methods in favour of natural practices as a way of life. An extreme case is the 'back-to-the-landers'. These urban immigrants, generally retired people, seasonal workers, or full-time professionals, practice intensive and semi-subsistence agriculture on farms no larger than 20 acres (Jacob and Brinkerhoff, 1986). Their life style includes an explicit rejection of the high-consumption/high-technology modern culture. Here, ideology is the key variable in determining technology adoption. Income from agriculture is a weak predictor, as most back-to-the-landers earn off-farm income as a primary source. Another group related to traditional/natural methods of farming is the Amish Society. Their methods are ruled by secular socio-religious considerations, although the new generations have accepted a blend of old with new ideas.[3] The Amish Society and back-to-the-landers share the rejection of some modern society values and can be taken as examples of how non-economical variables - ideology and religion - can explain adoption of sustainable technologies.

The literature presents other non-economic factors, such as sympathy for the ecological approach and concern about health (consumer, livestock, and farmer family health), affecting the farmer's technological choice. However, economic factors are not necessarily irrelevant in the decision process. Although the values and beliefs of ideological and religious groups may reduce the range of alternatives they face, most farmers hesitate in considering adoption of sustainable practices if they imply income reduction. Beus and Dunlap (1991), who measured farmers' basic beliefs and values, found strong evidence to distinguish three major groups in Washington State, USA. Alternative farmers were more interested in ecological and long term results than in productivity, efficiency and profit maximization. Adherents of conventional agriculture took an opposite perspective. An intermediary position was identified in a statewide farmers sample. In the alternative group, more philosophy-oriented subgroups, such as Permaculture adherents, were more motivated by non-economic considerations than certified organic farmers, who tended to put more importance on economic values.

The above discussion may suggest a dichotomy surrounds agricultural technologies. It is assumed environmental practices are designed specifically to protect natural resources, while conventional, or commercial, ones are conceived to increase farm economic efficiency. It also suggests environmental practices are unprofitable (or less profitable), while conventional ones are profitable (or more profitable) (Nowak, 1987). Assuming this dichotomy is valid, and considering that the two types of technology are adopted in the real world, one may infer that adopters of environmental practices are not profit-maximizers. Pampel and van Es (1977) followed this direction. They extracted from the literature three basic explanations of adoptive behaviour. The first is the 'psychological innovativeness' explanation, in which the willingness to change and try new ideas are the major forces, mainly for earlier adopters. There is only secondary interest in profit and environmental impact. The second explanation goes in the opposite direction: profit is the primary goal in adopting an innovation. This view is close to the assumptions in neoclassical economics, where the work of Griliches (1957) on hybrid corn is an example. Finally, the third explanation relies on farming as way of life. Some farmers adopt an innovation aiming for the satisfaction provided by the rural life style rather than business success. The psychological innovativeness reasoning can be used to predict both environmental and commercial practices. However, in terms of the other two explanations, profit orientation predicts adoption of commercial technologies, while farming orientation is the best predictor for environmental practices.

Two basic criticisms can be levelled at the above approach. First, the conflict between environmental and commercial innovations may not exist in cases where environmentally sound practices offer economic advantages. For instance, evidence suggests that minimum tillage and deep-rooted perennial species, which are relatively widespread, have been adopted for economic reasons rather than for their environmental benefits (Vanclay and Lawrence, 1994). Second, it can lead to the false conclusion that many agricultural producers do not adopt sustainable technologies simply because they are exclusively profit-oriented. Actually, as some conservation technologies are considered costly in the short term, economic constraints, such as insufficient capital or credit, can be the real barriers to adoption. Although a farmer may be concerned about the environment, these constraints may limit adoption (Heffernan and Green, 1986).

Barriers to adoption

Vanclay and Lawrence (1994) suggested that farmers have a rational basis for their reluctance to adopt environmental innovations. Although non-adoption can be explained by the specificity of commodities, environments and innovations being introduced, in the case of environmental conservation, additional obstacles can be found:

1 There is resistance to many environmental management practices because they are complex and require detailed understanding of physical process.
2 Given their whole farm approach, environmental innovations are generally non-divisible. Partial adoption, which is viewed as a form of trial, is difficult and farmers must be totally committed to the new practices before adoption.
3 Environmental strategies require great changes in agricultural practices, which may not be compatible with farm and personal objectives at a certain point in time.
4 Although environmental practices provide economic benefits for the society as a whole, they may not provide economic advantages for the individual farmer. Besides, conventional innovations, which have relatively short term benefits take precedence over environmental ones, which are more likely to have long term benefits.
5 Environmental innovations are particularly risky because not only are capital resources expenses involved, but also the production of one whole season can be lost. The new practices will only be adopted if the level of uncertainty about future benefits is low.
6 New technologies, especially those related to sustainability, are not free of debate. If the level of conflicting information is high, as is common for environmental innovations, non-adoption may be the appropriate strategy.
7 Some environmental innovations may require *capital outlay* in the form of machinery, seeds, and earthworks. Besides, income losses are expected until the new system is established. Many farmers cannot afford these costs,

while others in financial crisis, are unwilling to take any risk.

8 The knowledge basis of an individual farmer may not be adequate for the new practices. Many recommended strategies require technical knowledge of cropping systems, physical processes, chemicals, etc. An adopter must be highly motivated to acquire the necessary technical expertise.

9 Many environmental practices reduce farmers' flexibility in the sense that they would be restricted to a range of crops and rotations. In a world of fluctuating prices, they prefer to maintain the flexibility to respond to market signs.

10 Lack of physical and social infrastructure in the region can also create a barrier to adoption. Many commodities are tied to systems or marketing facilities that may not exist in the region. Additionally, the community may be reluctant to accept ideas that are different from the current local knowledge. Wide-scale adoption may wait until interest in the innovation has sufficiently grown.

11 The media has been presenting land degradation through cases of extreme consequences. This dramatic form of presentation is counterproductive because most farmers do not perceive themselves to be in those situations and, consequently, believe they do not have the problem. Moreover, those who identify their condition with the degree of degradation shown by the media, adopt a fatalistic attitude rather than taking remedial action.

Padel (1994) has also identified barriers to the conversion to organic farming. However, in contrast to Vanclay and Laurence (1994), she concluded these barriers did not imply general non-adoption, but rather a slow speed of diffusion. In Europe, the diffusion level of organic farming has been increasing since the second half of the 1980s, although it is still low.[4] Padel suggested organic farmers would fall under the category of 'innovators' and match Rogers and Shoemaker's (1971) characterisation with regard to higher level of education, relative youth, and problems related with social acceptance. Their smaller size and less commercial orientation, however, contradict the theory, although in Germany and Britain the average size of organic farms has been increasing, suggesting the financial motive has become more important and some barriers have been broken. Given the European problems with surplus production, environmental damage caused by agriculture and the greater concern about the effects of chemicals, organic farming has achieved a higher status and later adopters have been better accepted by local communities (Padel, 1994; Padel and Lampkin, 1994).

According to the diffusion model (Rogers and Shoemaker, 1971), information plays a fundamental role in adoption behaviour. However, if farmers suffer economic constraints, such as limited access to capital and land resources, they will be unable to adopt. Napier et al. (1984), using a sample of Ohio farms, found economic constraints were better predictors of adoption of soil conservation practices than diffusion-type variables, such as formal education and contact with different informational sources.

The profitability of sustainable technologies

Farmers may consider profitability an important criterion in the decision to adopt sustainable technologies. Relative prices of commodities and inputs, as well as interest rates and availability of credit, can significantly affect the decision. Particular environmental conditions, such as weather and soil quality, which can vary from one farm to another in the same region, are also important in determining adaptability and profit. Government policies, such as commodity programmes, have considerable influence on farmers' decisions, and can be especially relevant to decreasing the relative profitability of sustainable technologies. Alternatively, adoption can be favoured by environmental policies, such as a pesticides ban.

When a new technology replaces an old one, the time necessary to create new conditions or adapt the environment and production routines (specifically, in learning and acquisition of the necessary skills) can result in considerable cost to the firm. For sustainable technologies, the physical environment plays an additional role in the transition. Natural resources - soil and related biota - may have been degraded by years of constant use of chemicals. Soil quality may take time to be restored and, consequently, low yields can be obtained in the first years of the 'conversion period'. Income is expected to be reduced and this fact should be interpreted as an initial investment to be amortized in the future, when productivity is expected to increase. Income can also be initially reduced by a change in the production mix. Rotation with legumes, for instance, may lead to a reduction in cash crop production. These initial costs and uncertainty about future effectiveness generally induce farmers to take a gradual approach to adoption. Therefore, management skill, which is an important component of this kind of technology, plays a fundamental role, mainly during the first years after adoption. In the long run, risk-averse farmers, seeking more steady income, may be better-off in adopting sustainable practices.

Two basic methods have been used to assess profitability of sustainable technologies: enterprise or component analyses and whole-farm analyses (Madden and Dobbs, 1990). Enterprise budgets for crops and livestock operations are simple and present a good view of cost composition and profitability. They can be complete - including fixed costs, variable costs, and returns - or partial, comprising only variable costs and returns associated with farming practices. Although practical and easy to understand, enterprise budgets refer to one or a limited number of activities and ignore cross-effects. Sustainable farming systems rely on a holistic strategy, suggesting whole-farm analyses as a more appropriate approach. Optimization/simulation models may be used to produce a sensitivity analysis of whole-farm economic performance, instead of a partial view of a single activity. Data are not easy to obtain and the method has to rely on diverse information sources, such as experiment station trials, cooperating farmers, farm surveys, and case studies. A combination of these has been used in more sophisticated simulation models to achieve a more realistic view.

Lockeretz et al.(1981) used both component and whole-farm analysis to assess profitability of organic farming in the Corn Belt. The study was based on a multi-year comparison between conventional and commercial organic farmers.[5] The sample provides a two-period data set on the following variables: yields; application of fertilizers, manure, and other materials; tillage, cultivation and harvesting operations; and seeding rates and varieties used. For the 1974-1976 period, a set of 14 organic crop-livestock farmers in five State regions generated information through personal interviews and mailed surveys. A nearby conventional farm was selected for each organic farm generating a comparable data set. For the 1977-1978 period, a similar survey was conducted on 23 organic farms, but data on conventional ones were obtained from three different sources: country-level statistical reports, crop production budgets, and the 1974 Census of Agriculture. Whole-farm analysis showed the organic group average net returns were not very different from the conventional group: small positive and negative differences were found in alternate years. Organic farms per area operating expenses were the lowest, but offset by 6 to 13 percent lower production values. The greatest difference in profitability appeared in 1978, when conventional farms' net return was 17 percent superior. During that year, in contrast with the previous, growing conditions were favourable, suggesting that the economic advantage of chemical agriculture had relied on good weather. Under poorer weather conditions, organic practices matched, or even exceeded, yields from conventional agriculture. Component analysis revealed that yields of corn, soybeans, and wheat were respectively 10%, 5%, and 25% lower on organic farms, while for oats and hay, which were less chemical intensive, yields were equal in both farm systems.

Goldstein and Young (1987) used budget analysis to compare a conventional and a alternative farming system in Washington-Idaho Palouse, USA. In the past, sweetclover or alfalfa were employed in the region in rotation with winter wheat, barley, and peas. The system provided a solution for problems with disease, weeds and nitrogen fixation. During the 1950s and 1960s, it was replaced by methods based on intensive use of low cost fertilizers and pesticides but, as the conditions changed (higher chemical prices, increasing disease problems, and greater concern with the environment), interest in legume rotation reappeared in the 1980s. An alternative low-input system, called the perpetual-alternative-legume-system (PALS) was developed to substitute for the local chemical intensive four-year cereal rotation.[6] Economic comparison between the two systems revealed that the conventional one presented advantages when combined with high government support prices and the assumption of high yields. However, when market prices were considered, PALS produced higher profitability than the conventional rotation under both high and low yield assumptions. Also, price sensitivity analysis showed that PALS was economically advantageous under conditions of low wheat prices. As PALS annual production cost was 56 percent lower, farmers who adopted it, could additionally reduce risk and financial requirements.

A more complex nonlinear programming model with sensitivity analysis was used to establish profitable agricultural practices for Richmond County, Virginia, USA (Diebel et al., 1993). The model incorporated thirty-four production activities and considered four rotation systems. Pesticide/nutrient application rates were allowed to vary across this range, which comprised conventional and low-input agricultural methods. Commodity programmes and set-aside programmes were also included in the model. The results from an unrestricted scenario revealed that the two year corn/small grains-double-cropped soybean rotation, using organic fertilizer (poultry litter), was the profit-maximizing activity. As this outcome did not match the local production, a combination of penalties was imposed on litter price, on the yields of some activities using litter as a source of nitrogen, and on labour requirements for organic and low-pesticide activities. The penalties increased until organic methods were eliminated, although at that point low-pesticide activities were still accepted. The approach provided insights on possible economic barriers to adoption of low-input agriculture: cost and availability of organic fertilizer; high labour requirements; and low yields.

Most studies fail to consider transitional costs, on which data are not generally available. Hanson et al. (1990) presented a whole-farm analysis, which incorporated information from a nine-year Farming System Experiment held by the Rodale Research Center, Pennsylvania. Actual input data covering the period for a complete biological transition to low-input agriculture was provided by the Center. Prices and costs from 1981 to 1989, and also the effect of government programmes, were included in the model, giving a closer approximation to the real world than is observed in other studies. Versions of a conventional system and a low-input system were created, based on this experiment, resulting in four scenarios for a representative grain farmer in the Mid-Atlantic Region, USA: (1) a conventional grain farm, (2) a conventional grain farm participating in government feed grain programmes, (3) a low-input cash grain farm, and (4) a low-input cash grain farm participating in feed grain programmes.[7] The nine-year analysis found (2), the conventional farm participating in the government programmes, to be the most profitable scenario. This was followed by (4), the low-input farm participating in governmental programmes. But, model (4) was not replicable in real-life as it was created ignoring conflicts generated by the cross-compliance provisions. Third and fourth positions were occupied by the conventional and low-input farm scenarios, respectively.

In the above model, the cost of transition comprised costs related to the waiting time necessary for soil quality recovery and other special requirements, such as specific equipment purchases. A reduction in output prices was also considered, given the possibility that widespread adoption could depress the low-chemical food market. During the first four years, the annual average profits calculated for the low-input farm scenarios were lower than those calculated for the conventional ones, but, in the following years, this position was reversed. This outcome suggested that transitional costs could be amortized over the years that followed the biological transition. Thus, profits from low-input systems may match or exceed profits from conventional systems in a long term analysis. In other words, the 'economic transition' time may be longer than the biological transition time. The long-term analysis also allowed a comparison of the financial risks associated with each system. There was less variation in profits over the years in low-input systems than in conventional ones, suggesting they are the better choice for risk-averse farmers.[8]

In most studies profitability is evaluated at conventional prices. The market structure, however, can favour low-input products, as consumers may be willing to buy chemical-free food at a premium price. This is specially true when demand exceeds supply for such products, and where consumers are well-informed and certification standards have high credibility. In several countries, most organic products obtain premium prices, making organic methods' profitability comparable to that of conventional farming.

Organic producers in the Northeast of USA have received a 10% to 50% premium over conventional prices. However, many certified farmers in the country have no access to organic markets or, for non-economic reasons, intentionally avoid premium prices (Anderson, 1994). Similarly, in Australia and Canada premium prices are not widely available (Henning, 1994; Wynen, 1994). In those three countries, many farmers consider their organic production profitable, even at conventional prices but the same situation is not observed in Northern Europe (Britain, Germany, Denmark and Switzerland), where premium prices for most products are extremely important to the profitability of organic producers relative to conventional ones (Padel and Lampkin, 1994). Given the already high-intensive conventional agriculture, organic yields in those countries are generally lower.[9] Although variable costs are lower due to the reduction in agrichemicals, they are not low enough to compensate for lower physical productivity and higher labour costs. Premium prices are an essential requirement to avoid income losses.

Information sources

It was suggested in chapter 5 that information plays a fundamental role in the adoption process. In most countries, information on agricultural innovation has been delivered by government extension agencies and, for conventional technologies, integration into local information and assistance networks is very important. Adopters of sustainable technologies, however, generally have different information sources. They rely less on extension services and more on books, magazines and other printed materials, neighbours, group meetings, consultants, and personal contacts (Anderson, 1994; Thomas et al., 1990). Extension can still have some importance where official conservation programmes are involved, such as IPM and some widespread soil conservation practices.[10] This is not common, however, for more radical changes towards total elimination of chemicals. For example, Padel and Lampkin (1994) point out that extension agencies can lead to misconceptions concerning organic practices, yield expectations, financial performance and risk, which help to create barriers to conversion. They point out that organic farmers have access to information only through non-traditional sources.

Younger and better educated farmers have more contact with information sources than other farmers (Nowak, 1987; Rogers, 1983). Better education is particularly important for information processing and use of sophisticated management techniques (Anosike and Coughenour, 1990). One can expect those characteristics to be related to adoption of sustainable practices, mainly where there is no formal channel for information delivery and farmers have to search for alternative sources.

Another source of information is the extension service provided by non-governmental organizations (NGOs). They have become particularly important in Third World countries, where poverty and environmental degradation have occurred either with or without the introduction of Green Revolution technologies. In many areas, the introduction of conventional systems has not eliminated poverty and official agencies have lost legitimacy, creating a vacuum for NGOs to fill.[11] Where conventional technologies have not reached particular groups of farmers, especially small farmers, and traditional practices are not sustainable, NGOs have delivered information on low-input farming systems as a way of alleviating poverty and preventing land degradation (Reijntzes et al., 1992).[12]

Farm and farmer's characteristics

Empirical studies have shown that some specific characteristics of farms and farmers have favoured adoption of sustainable technologies. The role played by farm size, labour availability, farm environmental features, tenure, education and farming experience is reviewed here. The findings must be interpreted with care, as they are related to the level of diffusion already reached by sustainable practices. Diffusion is a dynamic process and some variables may change over time.

Farm size

The discussion of farm size and sustainable agriculture relies on the body of literature which underpins the agrarian environmentalism debate. The growing scale of modern agriculture has been criticized because of its socioeconomic and environmental impact. While the economic viability of small family farms has been undermined by lack of capital and unavailability of credit, large farms have received relatively more support from governments. From their favoured economic position, large farms could adopt expensive and ecologically unsound technologies.

The massive resource investment required by most Green Revolution technologies has inhibited the adoption of conservation methods. Large machinery, for example, compels farmers to become more specialized and reduces the economic feasibility of more holistic approaches. On the one hand, large farms have become relatively highly capitalized; on the other, they have faced increasing debt, reduced the number of cash crops, and assumed high risk positions. They adopted short-term profit-maximization behaviour, leaving little opportunities for environmental practices (Buttel et al., 1981). Critics of modern agriculture suggest that large farms have been put in an economic position that is incompatible with diversification (Oelhaf, 1978).

Anosike and Coughenour (1990), however, offer an opposite view and suggest that diversification is a risk-reduction strategy of large farmers. If specialization implies substantial economies of size, the farmer may face a tradeoff between profit and risk. If large farms have relatively higher variable cost, they confront greater risk and consequently tend to adopt a more favourable attitude to diversification. Thus, if high risk is a condition faced by large farms, diversification may increase with farm size. This reasoning, however, does not consider that risk can be reduced by external forces, such as government programmes, which may lead farmers in the opposite direction. Additionally, assuming that small farmers have limited scale economies, they do not face a tradeoff between profit (associated with specialization) and diversification. As a result, they may be more inclined to undertake diversification.

Buttel et al. (1981) found evidence to support an inverse relationship between farm scale, measured as gross income,[13] and farmers' environmental attitudes.[14] However, there was the caveat that they could not find significant correlation between gross farm income and concern about soil-erosion. Heffernan and Green (1986) showed that the larger the soil-erosion potential the larger the soil loss and suggested an alternative view about scale and environmental conservation. Large and capitalized farms, because they are located in the best lands, have relatively low soil-erosion potential. Small farmers tend to locate in affordable marginal areas, with higher soil-erosion potential. As a result, there are more soil-erosion problems on small farms than on large and capitalized ones, a conclusion which supports a negative correlation between scale and environmental degradation. As large farms have greater access to financial resources, they have more flexibility to allocate funds to conservation, while economic constraints limit the adoption of environmentally sound technologies by small farmers. For the same reason, the latter may have less access to more and better information. This is the position taken by Nowak:

> operators of larger-scale farms should have more flexibility in their decision making, greater access to discretionary resources, more opportunity to use new practices on a trial basis, and more ability to deal with risk and uncertainty associated with innovations (Nowak, 1987, p.211).

He found that number of acres operated was significantly and positively related to an index of adoption conservation practices.[15]

The role played by farm size on the adoption of sustainable technologies depends on technical, economic and institutional arrangements. Significant correlation may exist between farm size and other explanatory variables, such as credit, capital constraints, participation in government programmes, large machinery use, debt, information, and soil quality. It is especially difficult to isolate farm size from these variables in order to measure the pure effect of scale on the adoption decision.

Labour availability

The introduction of chemical and mechanical based technologies in agriculture has resulted in substantial labour displacement. In many countries, agricultural employment structure has been changing towards greater use of seasonal work and reduction of family labour, a farm-based resource. For some labour-intensive commodities, workforce demand has been particularly high during peak production periods, when seasonal hired labour is sometimes brought in from the outside community. Sustainable farming practices - as they rely on crop rotation, diversification, management, on-farm research, and reduction of chemicals - are expected to be more labour intensive than conventional technologies. Also, labour use is expected to be better distributed in time, resulting in a more sensible use of the family workforce during the year.

Pfeffer (1992) has studied perceived labour and production constraints on the adoption of practices that reduce chemical inputs. With a sample from southern New Jersey, he found most farmers agree that 'it is hard to reduce chemical inputs because additional labor in general is hard to find, skilled labor that can machine-cultivate crops is difficult to find, and their own labor inputs would have to increase' (Pfeffer, 1992, p.353). Surprisingly, it was also found that farmers who exclusively use family workers are more concerned about labour supply than farmers who rely on hired labour. The latter would have easier access to networks and sources of additional workers than the former. Generalizations, however, can be inappropriate, given that the conditions regulating elasticity of labour supply are usually locally determined. Local levels of unemployment can be high, or rural-urban migration can be in progress, both of which affect labour availability. For example, Diebel et al. (1993) found low-input and organic systems in Richmond County, Virginia were insensitive to labour requirements. Based on the existence of labour supply surplus in the County, the authors made a pooled labour supply assumption in a nonlinear programming model.

In northern Europe, labour use on organic farms is higher than on comparable conventional ones (Padel and Lampkin, 1994). This is explained not only by specific technical requirements, but also by the enterprise mix of this farming system. The production share of vegetables and root crops, which are more labour intensive, is increased after conversion. Outside the production level, the need to obtain organic premium prices also leads to increased demand for labour; with more hours of work being devoted to marketing development and processing activities.

Off-farm income can be useful in providing financial resources for the conversion to sustainable agriculture. However, if the additional funds come from off-farm work, labour availability can be limited. In a sample of Kentucky farmers, for instance, diversification was found to be negatively affected by off-farm work (Anosike and Coughenour, 1990). Here, the development of multiple enterprises was constrained by the reduced number of hours available for on-farm work.

Education and farming experience

The level of education is an important factor in the adoption of sustainable practices. Education is related not only to the ability to obtain and process information, but also to the use of more sophisticated management techniques. It has also been recognized that ecological methods require a higher level of expertise than conventional practices (Lockeretz, 1989). Thus, one can hypothesize that farmer's education and experience are important personal characteristics in the adoption of sustainable practices.

Empirical studies have revealed that low-input and organic farmers are better educated than their conventional counterparts. However, they are younger and have less experience in agriculture (Anderson, 1994; D'Souza et al., 1993; Padel and Lampkin, 1994; Henning, 1994; Thomas et al., 1990). In fact, although greater experience - measured by age or years of farming experience - is an important element to explain the level of management ability, it has some negative effects on the adoption of sustainable practices. Older farmers may be less energetic (Anosike and Coughenour, 1990) and have a shorter planning horizon (Rahm and Huffman, 1984), while younger farmers are more attracted to novelties and more likely to be early adopters (D'Souza et al., 1993).

Farm environmental characteristics

The probability of success of an agricultural technology depends on its appropriateness and compatibility with a farms' ecological conditions. For instance, many high yielding varieties were designed to be reliable under good irrigation systems, and some mechanical innovations cannot be used for hill farming. Soil types, terrain, topography, water supply, and climate are characteristics that vary from one region to another, and sometimes between farms in the same region. The farmer's adoption decision is affected by the particular physical environment. It is reasonable to suppose that adoption of sustainable technologies is also influenced by the suitability of the innovation to specific physical conditions.

There are few studies on sustainable technology adoption in which ecological variables have been thoroughly evaluated. Data are difficult to obtain, making a complete characterization a difficult task. In Great Britain, organic farming is predominantly located in areas where cropping is feasible; very few farms have been found in areas where climatic factors restrict activities to livestock only (Lampkin, 1994). Diversification in USA farms is greater where local environmental variation is large (Anosike and Coghenour, 1990), and the probability of reduced tillage adoption has been found higher where soils are rolling lighter and better drained (Rahm and Huffman, 1984).

Tenure

Renters are supposed to have a shorter planning horizon than owners. Depending on the tenurial agreement, they have greater uncertainty about, or may not expect to receive, benefits from soil improvement. Therefore, owners tend to invest more in conservation technology than renters (Nowak, 1987). Additionally, renters have greater risk aversion, which inhibits adoption of new technologies (Rahm and Huffman, 1984).

Empirical studies, however, have presented conflicting evidence. The relation between tenure and reduced tillage practice adoption has been found not significant (Nowak, 1987; Rahm and Huffman, 1984). In the USA, the percentage of owners among organic farmers is greater than among conventional ones (Anderson, 1994). Anosike and Coughenour (1990) found diversification decreases with landownership. In fact, the same conflict has been found in studies on adoption of Green Revolution technologies (Feder et al., 1985). A possible explanation is the fact that tenure may be indirectly related to access to credit, input and product markets, and information. Also, the terms of tenurial agreement, which is rarely specified in empirical studies, may, or may not, persuade owners/renters to embrace land conservation practices.

The role of agricultural and environmental policies

In many countries different policy instruments have been used to support agricultural production. Deficiency payments, import levies and export subsidies are the most important approaches. It has been recognized that some of these policies are incompatible with environmental targets (National Research Council, 1989; OECD, 1989; and OECD, 1993). In the USA, commodity programmes have encouraged farmers to use an excessive amount of external inputs in order to achieve higher yields and maximize government payments. Subsidized crop production has expanded into marginal or poorly suited land through intensive use of chemicals. In the EC, high levels of subsidies have also led to excessive intensification. Commodity programmes, through different levels of support, have differentiated crop and livestock activities, inducing farmers to reduce enterprise mix. Some crops, such as maize in the USA and sugar beet in the EC, have been highly supported while root crops have received no support. Consequently, monoculture, which has presented economic advantages over diversification, has caused severe environmental stress. Research and development has followed economic signals, and some privileged activities have received higher incentives than those promoting environmental values.

The National Research Council (1989) reported that two components of USA commodity policy have had a negative impact on the adoption of alternative practices in agriculture: base acre requirements and cross-compliance provisions. The benefits received by farmers are related to the crop acreage planted and yield obtained in the previous five years.[16] Under this policy, farmers were compelled to maintain the crop acreage planted in order to maximize benefits. A reduction in the planting of a particular crop in a certain year would lessen the eligible acreage base, resulting in benefit losses in that and subsequent years. The 1985 Food Security Act introduced financial penalties to the expansion of programme crop base acreage. A cross-compliance provision established that to receive the benefits from a settled crop acreage base, the farmer could not increase the acreage base for any other crop programme. As a result, adoption of rotation practices that would imply increase in other crop acreage would cause ineligibility for the programme and consequent financial penalty. The need to maintain base acres and the cross-compliance provision created economic barriers to widespread adoption of conservation practices. Only farmers outside the programmes, or those already diversified, would not be penalized by adopting alternative farming systems.

Lampkin and Padel (1994) have found similar barriers to the adoption of organic farming systems in Western Europe.

> Arable farms converting to organic production will tend to reduce the arable area and increase the area of grassland and the number of livestock, while livestock farms will tend to reduce livestock numbers and may introduce or expand arable production.
> Arable farmers converting now will, in addition to the other costs of convertion, lose eligibility for some arable area payments, without compensation, but can only access to some livestock premiums through

quota purchase. Livestock farmers converting will receive livestock payments on fewer animals, yet will not be entitled to arable acre payments for any new arable land introduced, although this may be offset by quota sales (Lampkin and Padel, 1994, p.446).

In recent years, an increasing effort has been made towards a better integration of agricultural and environmental policies. This task has been facilitated by the economic factors that led to the reform of the European Union's Common Agricultural Policy (CAP) and the 1993 GATT agreement.[17] Another force for change is the international effort towards trade liberalization. Although the 1993 GATT agreement was limited in the removal of agricultural trade barriers (Ingco, 1994), many countries are adjusting their agricultural polices towards a greater market orientation.

There is a gradual shift to reduced protection and an increasing interest in 'decoupled' forms of support that are not linked to production, including payments associated with the provision of environmental services (OECD, 1993). The effect of CAP reform on conservation culminated in the introduction of specific measures to encourage adoption of agricultural methods compatible with environmental protection and the maintenance of the countryside (EC, 1992). Member States should implement, in accordance with their particular needs, financial aid programmes, courses, training and demonstration projects for farmers who undertake:

(a) to reduce substantially their use of fertilizers and/or plant protection products, or to keep to the reductions already made, or to introduce or continue with organic farming methods;
(b) to change, by means other than those referred in (a), to more extensive forms of crop, including forage, production, or to maintain extensive production methods introduced in the past, or to convert arable land into extensive grassland;
(c) to reduce the proportion of sheep and cattle per forage area;
(d) to use other farming practices compatible with the requirements of protection of the environment and natural resources, as well as maintenance of the countryside and the landscape, or to rear animals of local breeds in danger of extinction;
(e) to ensure the upkeep of abandoned farmland or woodlands;
(f) to set aside farmland for at least 20 years with a view to its use for purposes connected with the environment, in particular for the establishment of biotope reserves or natural parks or for the protection of hydrological systems;

> (g) to manage land for public access and leisure activities. (EC, 1992, p.86-87)

In the Scandinavian countries, Austria, and Germany, farmers converting to organic agriculture receive financial aid from special schemes, which also comprise the development of extension, information and marketing services (Lampkin and Padel, 1994). France, UK, and Switzerland are also offering financial support for farmers, research and advisory services on organic farming (OECD, 1993).

Arable land set-aside policy has been applied in many countries. It was initially conceived to withdraw agricultural land from production and reduce market surpluses. However, in Canada and the USA a link has been established between set-aside and the conservation of land, mainly in areas with highly erodible soils. In Europe, where soil erosion causes less concern, environmental objectives are less important (OECD, 1993), although a link has been created in recent years.[18] There are two opposite views on the effect of set-aside on the adoption of environmentally friendly practices. On one hand,

> the EC set-aside scheme is seen as having a potential role in helping farmers to convert to organic production. The great majority of farmers participating in the scheme are choosing the fallow option which could provide an opportunity to build up organic soil fertility (OECD, 1993, p.61).

On the other hand, set-aside policies 'can create an incentive to increase yields on land still under cultivation' (OECD, 1994, p.74). The relevance of one or other of these views to explaining the effect of set-aside depends on specific schemes created for conservation. In many countries, increasing yields through further intensification may not be economically feasible, because extra charges on fertilizers and pesticides have been introduced to reduce agricultural pollution. It is reasonable to expect a more positive effect of set-aside on the adoption of sustainable technologies in countries where restrictions on agrichemical use are tighter.

In the USA, crop programmes have been changed to allow compatibility with environmental targets. More flexibility is now given to farmers in growing rotation crops without incurring loss of their base acres upon which programme payments are based (OECD, 1994). Additionally, cross-compliance provisions have been created to protect the environment (Russell and Fraser, 1993). Farmers must plan and implement erosion control and wetland management practices in order to remain eligible to receive support payments. An increasing number of countries are pursuing this type of policy. Financial assistance has also been provided to encourage the adoption of farming methods that integrate many farming activities into a management system.

The focus on sustainable production systems is growing more salient in government R&D programmes although public budgets have been cut in many countries. Research programmes have included the following themes: crop management systems (e.g. crop rotation systems, tillage, pesticide and fertilizer management), ecological parameters (e.g. soil chemistry and structure, hydrological and nutrient cycles), production and processing techniques, integrated production systems, animal waste management, comparative economic studies for conventional and alternative production, and genetic resources and biodiversity (OECD, 1994). More attention has also been given to information transfer. Japan and the Netherlands have stimulated, or created, demonstration projects. In many USA States the extension service is trying to develop programmes to link research with on-farm demonstrations.

Conclusion

This chapter provided a review of empirical studies of the factors affecting adoption of sustainable technologies in agriculture. It was seen that not only non-economic reasons but also economic considerations have led farmers to decide in favour of new approaches to agricultural production. Many of these factors are beyond the farmer's control but some have been changing in favour of a more sustainable approach. Although technical and economic barriers are still present, there is a shift towards integration of agricultural and environmental policies. More information is available, markets for environment-friendly products have developed, and later adopters have been socially better accepted. Moreover, many sustainable practices are becoming economically feasible. As economic and social barriers have been reduced, new entrants are receiving more positive signals than the earlier adopters.

Most empirical studies on sustainable technology adoption have failed to consider these changes. Statistical techniques (correlation analysis, multiple regression, and linear and probit/logit probability models) have been applied to discriminate between adopters and non-adopters but the analysis have been static. Two basic criticisms can be made of these approaches. First, it is rarely taken into account that farmers adopted an innovation at different points in time. The data available generally refer to the date of the cross-section survey instead of the time of adoption. Second, the influence of variables, such as prices, which may not change from one individual to another but change as time passes, is not evaluated. As indicated in the next chapter, duration analysis has been used in other fields to investigate the impact of time-varying factors. Given that agricultural and environmental policies, and market structure have been changing over time, and economic factors have assumed greater importance in the adoption of sustainable technologies, future research should consider the appropriateness of more dynamic approaches to decision-making modelling.

Notes

1 Even rural residents not engaged in non-extractive activities are expected to have similar attitudes because of their involvement with the rural culture. Rural residents have 'a greater tendency to hold utilitarian (as opposed to appreciative) attitudes towards the natural environment, when compared to urban residents' (Tremblay and Dunlap, 1978, p.476).

2 During the 1970's, financial advantages were associated with higher prices created by increased demand for organic food. Also, high chemical fertilizer prices and federal restrictions on cheaper pesticides induced a more favourable attitude to alternative practices (Oelhaf, 1978).

3 The mix of old and new technologies used by the new generation of Amish Society has been suggested to be similar to the academic proposal of low-input sustainable agriculture (Stineer et al., 1989). As off-farm income is not a predominant characteristic of this group, one can deduce that their greater dependence on agriculture allowed the new generation easier acceptability of some modern society technologies.

4 In 1993, organic methods were applied in less than 1% of most Western European countries' agricultural land. In Germany, where the area with organic agriculture is the largest in Europe, organic farmed land has increased from 29,000 hectares in 1985 to 197,000 hectares in 1993, although the diffusion rate in this last year was only 1.11% (Padel, 1994).

5 In this study, organic farmers 'were neither older farmers who never got around to adopting modern chemicals and other modern methods nor members of any youthful counterculture movement.' Here, there is strong emphasis in characterize organic farmers as 'more like their neighbors who farmed conventionally than like the stereotyped organic farmers described by either supporters or critics' (Lockeretz et al., 1981, p.541).

6 The conventional system consisted of four-year winter wheat, spring barley, winter wheat, spring peas rotation, while the alternative system was a three-year rotation of spring peas/black medic, black medic, and winter wheat. Black medic is a biannual legume which is able to reseed itself for as long as 30 years (Madden and Dobbs, 1990) and competes little with other crops

(Goldestein and Young,1987).

7 The size of the representative farm was 750 acres of tillable cropland. The conventional system was a five-year rotation of corn, corn, soybeans, corn, and soybeans, while the low-input system was a rotation of small grain/forage legume, corn, small grain/forage legume, corn, and soybeans.

8 In organic agriculture, risk is reduced due to enterprise diversity. Lack of information or partial analysis can lead to the false perception that some alternative farming systems are riskier than conventional agriculture (Padel and Lampkin, 1994).

9 In the USA and Australia, where conventional agriculture is less intensive, conversion does not cause much, if any, decrease in physical productivity, resulting in relatively less impact on profitability (Lampkin and Padel, 1994). Henning (1994) reported that farmers in Canada were motivated to convert to organic methods for a variety of reasons, including the expectation of greater profits. As many farmers faced financial stress, the author concluded 'some farmers regard organic practices as a strategy for survival'.

10 A study of adoption of IPM practices among Texas cotton growers revealed two important sources of information: group meetings and contacts with extension and university professionals (Thomas et al., 1990).

11 Exceptions, however, can be found in some countries - Guatemala, Bangladesh, and Bolivia - where exceptionally it has been recognized that NGOs have worked together with government agencies (Biggs, 1990).

12 Buttel suggested that 'many environmentally-oriented NGOs, in fact, were previously social-justice oriented ones or are staffed by persons whose commitments are as much or more to social justice as to environmental conservation. Many small environmental and sustainable development-oriented NGOs, particularly those of Third World origins, self-consciously employ environmental claims as a calculated means of agitating for social justice goals' (Buttel, 1992, p.18).

13 The most commonly used indicator of farm scale is gross income, although the literature presents other variables, such as net income, acres farmed, acres owned, hired labour.

14 Three environmental attitudinal variables: support for regulation of industrial pollution; concern with pollution from agricultural chemicals; and concern with soil-erosion, were regressed on socioeconomic variables (Buttel et al., 1981).

15 The same study contained evidence contrary to Heffernan and Green's (1986) findings that larger farmers own the highest quality land.

16 In 1985, base acreage yields were frozen at the 1981 to 1985 average (National Research Council, 1989).

17 Given the pressure of tied budgets, support programmes and overproduction has caused concern. Some authors have argued that the major reason for CAP reform has been attributed to the fiscal cost associated with agricultural policies (Bonanno, 1991). This view is put in question by the fact that a system of direct income payments was introduced to compensate income losses. 'Further, the MacSharry reforms were enacted in parallel to the Uruguay Round of GATT negotiations. The CAP was the subject of international negotiation' (Kay, 1995, p.3).

18 The introduction or maintenance of sustainable practices has been encouraged under the European Environmentally Sensitive Areas Scheme. 'ESA designation is oriented towards farming systems, which are environmentally friendly, rather than to specifically defined conservation objectives. In the UK these powers have been used to designate coherent areas of important landscape and wildlife interest which were perceived to be under threat from either more intensive farming or from neglect arising from the weak economic state of the local farming system' (Colman et al., 1992, p.32).

7 Two econometric techniques for the study of technology adoption: probit/logit models and duration analysis

Introduction

This chapter provides a review of two econometric techniques that have been employed in empirical studies of technology adoption. The first, probit and logit models, were widely used during the 1980's, while the second, duration analysis, has only recently been applied to the analysis of adoption and diffusion of innovation. The main objective here is to depict the econometric basis of the empirical analysis reported in chapter 8.

Probit and logit models

It is not uncommon to find econometric models in which the dependent variable, say y, assumes discrete rather than continuous values. In technology adoption, for instance, the decision to adopt can be modelled by regressing a dummy variable for the individual status (1 - adopter, 0 - nonadopter) on a set of explanatory factors. As will be shown, the conventional regression model is inappropriate for this kind of problem and logit and probit regressions are a better choice.

Empirical studies on agricultural technology adoption have used several econometric methods to draw conclusions from available data. The simplest method is correlation analysis. This technique is capable of generating qualitative information about the interrelationships of distinct factors, although it fails to provide a quantitative measure of their importance. To supply this need, ordinary least square regressions have been used. Generally, a dummy variable (1-adoption, 0-nonadoption) is regressed on a set of explanatory variables, giving the model:

$$y_i = E(y_i|x_i) + \mu_i = \beta' x_i + \mu_i ,$$

where x_i is a set of explanatory variables, β is a vector of parameters, and μ_i is a randon error with $E(\mu_i) = 0$. The calculated value of y from the estimated regression $\hat{y}_i = \hat{\beta}' x_i$ is supposed to represent the probability of adoption, given the explanatory variable values of the vector x. This linear probability model offers advantages over simple correlations, although it still suffers from some shortcomings, as will be shown below.

As y assumes only two values, the residual is a discrete random variable with values $-\beta' x_i$ and $1-\beta' x_i$. Their respective probabilities are

$$Prob(1-\beta' x_i) = Prob(y_i=1|x_i) = E(y_i|x_i) = \beta' x_i$$

and

$$Prob(-\beta' x_i) = Prob(y_i=0|x_i) = 1 - Prob(y_i=1|x_i) = 1 - \beta' x_i .$$

With these expressions, the following results are obtained:

$$E(\mu_i) = (1 - \beta' x_i)\beta' x_i - \beta' x_i(1 - \beta' x_i) = 0$$

and

$$Var(\mu_i) = E(\mu_i^2) - [E(\mu_i)]^2$$

$$= (1 - \beta' x_i)^2 \beta' x_i + (\beta' x_i)^2 (1 - \beta' x_i)$$

$$= (\beta' x_i) (1 - \beta' x_i),$$

which reveals the existence of heteroscedasticity. Consequently, OLS estimates of β are not efficient. The problem can be solved by two-step weighted least squares, but, as the μ_i s are not normally distributed, the estimates will not be fully efficient. However, the major problem of linear probability models comes from the fact that, although the observed values of y are 0 and 1, the calculated values can actually lie outside these limits. Thus, the model fails to predict sensible probabilities.

A solution to the above deficiency can be given by probit and logit models. There are several possible specifications in this category of models and, generally, they can be distinguished according to the number and characteristics of possible dependent variables. Regarding the number of dependent variables, three basic subcategories may be defined: univariate, bivariate and multivariate. A further classification is given by the number of values which the dependent variable can assume. In cases where only two values are undertaken, the models are said to be binomial. If more than two values are admitted, a multinomial model is obtained. Additionally, the latter can be either a categorial one, if y is a dummy for different categories, or noncategorial. Finally, a categorial dependent variable can be ordered, unordered or sequential.[1] The simplest models are the binomial univariate probit/logit. In economics, they were initially used in consumer decision modelling. Later, adoption and diffusion empirical studies widely employed this approach.

Let y^* be a 'latent' or unobserved variable defined as

$$y_i^* = \beta' x_i + \mu_i. \tag{7.1}$$

The observed values of y are still 0 or 1, but they are now supposed to be related to y^*_i as follow:

$$\begin{aligned} y &= 1 \quad \textit{if} \quad y_i^* > 0 \\ y &= 0 \quad \textit{otherwise.} \end{aligned} \tag{7.2}$$

Here, y^* is not constrained and can take a continous distribution. From 7.1 and 7.2, we have

$$Prob(y_i = 1) = Prob(y_i^* > 0) = Prob(\mu_i > -\beta' x_i) = 1 - F(-\beta' x_i),$$

where F is the cumulative distribution function for μ. If a symmetric distribution is assumed,

$$Prob(\mu_i > -\beta' x_i) = Prob(\mu_i < \beta' x_i) = F(\beta' x_i).$$

For n independent observations and assuming a symmetric distribution for μ, a likelihood function is obtained:

$$L = \prod_{i=1}^{n} [Prob(y_i = 0)]^{(1-y_i)} \; Prob(y_i = 1)^{y_i}$$

$$= \prod_{i=1}^{n} [1 - Prob(y_i = 1)]^{(1-y_i)} \; Prob(y_i = 1)^{y_i}$$

$$= \prod_{i=1}^{n} [(1 - F(\beta' x_i)]^{(1-y_i)} \; F(\beta' x_i)^{y_i}.$$

Estimation of parameters follows maximum-likelihood procedures.

Assumptions about the functional form of F result in different models. For a normally distributed μ, the probit model is obtained, giving

$$Prob(y_i = 1) = \int_{-\infty}^{\beta' x_i} \phi(t)\, dt = \Phi(\beta' x_i), \qquad 7.3$$

where ϕ and Φ are the density and cumulative density functions of the normal distribution, respectively. If a logistic function is assumed, we have

$$Prob(y_i = 1) = \frac{e^{\beta' x_i}}{1 + e^{\beta' x_i}}, \qquad 7.4$$

yielding the logit model.

The normal and the logistic differ only in the tail of the distribution, where the logistic is wider. In practice there is little to choose between them, unless the sample is very large.[2]

The estimated coefficients cannot be interpreted as marginal effects. Whatever the distribution of μ, we have

$$E(y) = Prob(y = 1|x) = F(\beta' x).$$

Hence,

$$\frac{\partial E(y)}{\partial x} = [\frac{dF(\beta' x)}{d(\beta' x)}]\ \beta = f(\beta' x)\ \beta, \qquad 7.5$$

where $f(.)$ is the probability density function. The marginal effects of parameters for the probit and logit models are:

$$\frac{\partial E(y)}{\partial x} = \phi(\beta' x)\ \beta$$

and

$$\frac{\partial E(y)}{\partial x} = \frac{e^{\beta' x}}{(1 + e^{\beta' x})^2}\ \beta\ ,$$

respectively. Generally, the model is interpreted at the mean values of the explanatory variables.

Specification tests

Tests of hypotheses about coefficients follow the standard approach for maximum-likelihood estimates. Standard errors from the information matrix can be used to perform t tests. For parametric restrictions, the trinity formed by Wald, Likelihood Ratio and Lagrange Multiplier tests is available. A common practice is to test the hypothesis that all coefficients except the constant are equal to zero. This gives some insight into the importance of the set of explanatory variables in the model.

In conventional regression models, R^2 provides a goodness of fit measure. Its counterpart here can be given by the Likelihood Ratio Index:

$$LRI = 1 - \frac{lnL}{lnL_0} ,$$

where L is the likelihood function evaluated at the full set of variables and L_0 is the same function but computed with only the constant term. If all of the coefficients are zero, LRI equals zero. However, as LRI does not reach the value 1, values between 0 and 1 cannot be interpreted in the same way as R^2.[3] Nevertheless, this test can be useful for comparing different model specifications.

Heteroscedasticity can be an important specification problem in probit and logit models, mainly because they are generally applied to cross-section data. A convenient formulation for a version with heteroscedasticity is:

$$y^* = \beta' + \varepsilon$$

$$Var[\varepsilon] = [\exp(\gamma' z)]^2,$$

where z is a vector of a sub-set of variables and γ a vector of parameters. The likelihood function will be

$$L = \prod_{i=1}^{n} [(1 - F (\frac{\beta' x_i}{\exp(\gamma' z_i)})]^{(1-y_i)} \ [F (\frac{\beta' x_i}{\exp(\gamma' z_i)})]^{y_i} .$$

With the estimates of parameters one can test for homocedasticity by examining the hypothesis that $\gamma = 0$. For this task, all of the three suggested tests apply.[4]

Finally, predictability of the models can be assessed through a 2 X 2 table comprising correct and incorrect predictions. In this case, a prediction rule is necessary. Greene (1992) uses the following:

$$\hat{y} = 1 \quad if \quad \hat{\beta}' x_i > 0, \ and \ 0 \ otherwise$$

A possible table format is exemplified in Table 7.1. The model's predictability is evaluated by examining the ratio given by the number of correctly predicted zeros and ones (a+d) over the number of observations (a+b+c+d). This ratio ranges from 0, the worst possible case, to 1, for perfect predictions.

Table 7.1
Logit/probit models predictability matrix

Actual (y)	Predicted ($\hat{y}$) *0*	*1*
0	a	b
1	c	d

a Number of correctly predicted zeros
b Number of incorrectly predicted ones
c Number of incorrectly predicted zeros
d Number of correctly predicted ones

Duration analysis

Duration analysis has a long tradition in biometrics and statistical engineering but only recently has it arisen in econometrics. Lancaster's (1979) study on unemployment appears to be the first to apply this technique in the social sciences. To introduce the idea, let us give some examples of problems to which duration analysis can be applied. In medical research it is illustrative to take the case of heart transplants. In this area, researchers are usually interested in knowing which variables affect the survival time of heart transplant recipients. Presumably, one of these variables is the survival time itself, given that the longer an individual stays alive after a successful heart transplant the higher the probability of dying in the next period of time. Also some donor-recipient variables, such as age, can affect the probability of death. Thus, the answer can be formulated by testing the explanatory power of different factors, including time, on the probability of dying at a certain moment. In unemployment analysis, researchers use the technique to investigate possible explanations for the length of time a person stays unemployed. The set of explanatory variables here comprises not only how long he or she has been out of a job, but also other factors, such as the rate of job offers received, the wages payable in those offers and the level of unemployment benefits that can be obtained. Personal and social characteristics, such as age, marital status and race, might also have some influence. Similarly, researchers on innovation adoption are interested in knowing what variables can affect the length of time a firm will wait to adopt a new technology from the time it becomes available. Firms' different waiting times may be explained by several factors, such as firm size, market concentration, adoption by rivals, profits, legislation, etc.. It is easy to see that, although the above examples emerge from different research areas, they are amenable to similar statistical treatment, and this is provided by duration analysis.

The main objective of this section is to present the basic concepts of duration analysis (also known as survival analysis). Our main concern is to cover only the material necessary to make understandable the validation and results of the method. A more comprehensive treatment can be found in the textbooks and articles referred to in the text. The presentation attempts to be as simple as possible, focusing on the models as they are found either in standard econometric computer packages or in adoption technology studies.[5]

The duration time variable

The examples presented in the introduction have a dependent variable which is the duration of a process, or alternatively, the length of time measured from the beginning of an event until its end: the survival duration of a heart transplant recipient, the length of time that a person stayed is unemployed, and the time a firm waits to adopt a new technology. Thus, two points in time have to be defined in order to obtain a value for the duration time: the entrance and the exit dates. The entrance date, or the beginning of a process, can be either directly observed at different points in calendar time or simply fixed at a point in time. Unemployment spells, for example, may start randomly in real time, as the entrance date normally differs from one individual to another. Alternatively, in technology adoption studies, it is common to set the initial point either at the time when the first adoption took place or, if the firm was created after that, at the time of its creation.

Obviously, the exit date, or the end of a process, has to be set at its terminal point; for example, at the time when a patient dies, a person accepts a job offer, or a firm adopts an innovation. Nevertheless, in practical terms the available data for social researchers are usually gathered by cross-section surveys. Some spells may have not been completed at the time of data collection. Some persons might still be unemployed or some firms may not have adopted the technology by that time. In other words, the end of some spells is unknown, although they might come in the future. For these cases the statistical procedure is to 'right-censor' the duration at the end of the observation period, that is at the time when the data were collected.[6] It is interesting to note that, in cases where the entrance date is set to be the same for all individuals, the exit calendar time coincides with the duration length.

The parametric estimation of the duration distribution

Probability theory plays a fundamental role in duration analysis. An easy starting point for understanding this matter is to reformulate the problems presented at the beginning of the chapter. Instead of focusing on the time length of a process, one can consider the probability of its end, or, as it is the same, the probability of transition to a new process. In the heart transplant case, one can ask about the probability of a person dying in the tenth year, given he or she has been a heart transplant recipient for nine years. In the unemployment study, the question might be: what is the probability of an individual getting a job in the sixth week, given he or she has been unemployed for five weeks? In the technology adoption study: what is the probability of a firm adopting a certain technology after time t, given it has not adopted it by that time? The answers are generated by the 'hazard function', defined as the probability of a spell of time to be completed after duration t, given that it lasts until t.

Let f(t) be a continuous probability distribution of a random variable T, where t, a realization of T, is the length of a spell. The corresponding cumulative density is given by

$$F(t) = \int_0^t f(s)\, ds = Prob\ (T \leq t). \qquad 7.6$$

Alternatively, the distribution of T can be expressed by

$$S(t) = 1 - F(t) = Prob\ (T \geq t), \qquad 7.7$$

which is the 'survival function', or the cumulative density turned around. S(t) gives the probability that a spell is of length at least t, that is, the probability that the random variable T is equal to or exceeds t. The hazard, as defined above, can be expressed by $Prob(t \leq T \leq t + \Delta \mid T \geq t)$, where Δ is the next short interval of time after t. The limiting value of this probability divided by Δ, when Δ tends to zero, gives the hazard function

$$h(t) = \lim_{\Delta \to 0} \frac{Prob\ (t \leq T \leq t + \Delta \mid T \geq t)}{\Delta}$$

$$= \lim_{\Delta \to 0} \frac{F(t+\Delta) - F(t)}{\Delta\ S(t)} \qquad 7.8$$

$$= \frac{f(t)}{S(t)},$$

which, in medical research for instance, is the 'instantaneous' death rate for an individual surviving to time t.

The cumulative density, survival and hazard functions are alternative and mathematically related functions through which the distribution of T can be expressed. Another useful function, as will be shown later, is the 'integrated hazard'

$$\Lambda(t) = \int_0^t h(t)\,dt = -\ln S(t). \qquad 7.9$$

The distribution of T can assume any parametric specification in a model of duration analysis. Table 7.2 shows the corresponding survival and hazard functions of the most common distributions of T. The choice of distribution may emerge from economic theory or be governed by convenience. The hazard for the exponential distribution, for example, is a constant, meaning that the conditional probability of 'failure', or change of state, in a given short interval does not depend on duration. For this reason it is called 'memoryless', that is, the passage of time does not affect its value. To allow for time dependence, one has to rely on other distributions, such as the Weibull, normal, logistic, Gompertz and gamma. The Weibull hazard, for instance, increases or decreases monotonically, depending on parameter p.[7]

Once the parametric distribution of T has been chosen, estimation of parameters follows maximum likelihood procedures. Assuming the duration of each individual, t_i, is independent of the others, the log-likelihood function for completed spells is

$$L(\theta) = \sum_{i=1}^{n} \ln f(t_i, \theta);$$

Table 7.2
Survival and hazard functions for some distributions*

Distribution	Survival Function	Hazard Function
Weibull	$\exp(-(\lambda t)^p)$	$\lambda p(\lambda t)^{p-1}$
Exponential	$\exp(-\lambda t)$	λ
Normal	$\Phi(-p \log(\lambda t))$	ϕ/Φ
Logistic	$1/(1+(\lambda t)^p)$	$\lambda p(\lambda t)^{p-1}/(1+(\lambda t)^p)$

* Formulations correspond to Greene (1992)

where $f(t_i,\theta)$ is the density function and θ is the parameter vector, which, for the moment, comprises only the parameters λ and p. In cases where censored observations are included, information on their exact durations is not available and the density function cannot be applied. However, we know that the duration of these observations is at least t_j. In other words, the survival function, $S(t_j,\theta)$, is available. Thus, the likelihood function becomes

$$L(\theta) = \sum_{i=1}^{n} d_i \ln f(t_i,\theta) + \sum_{i=1}^{n} (1-d_i) \ln S(t_i,\theta), \qquad 7.10$$

where d_k=1 if the kth spell is not censored and d_k= 0 if censored.

The log-likelihood function can also be formulated in terms of the hazard function. Taking the mathematical relation 7.8 and substituting in 7.10 gives

$$L(\theta) = \sum_{i=1}^{n} d_i \ln h(t,\theta) + \sum_{i=1}^{n} \ln S(t,\theta). \qquad 7.11$$

Alternatively, taking 7.9 and substituting in 7.10 gives

$$L(\theta) = \sum_{i=1}^{n} d_i \ln h(t,\theta) - \sum_{i=1}^{n} \Lambda(t,\theta). \qquad 7.12$$

Maximum likelihood procedures can be used to estimate the θ parameters.

Explanatory variables

In the above model, the estimated parameters, p and λ, define the 'shape' and the 'scale' of the distribution of T. Explanatory variables, or 'covariates', are not introduced. However, they will be included in the models of the next three sections. It will be seen that one advantage of duration analysis is the possibility of measuring not only the effect of variables that stay constant over time but also the effect of variables that change over time. Thus, it might be useful to classify covariates according to their relation with time. Some covariates, such as personal and social characteristics, do not change, or are assumed to be constant over time. Gender and race, for instance, are naturally supposed to be 'time invariant'. Lack of information can also lead the researcher to assume that some variables are constant over time. For example, the data available on firm size may only refer to the time of the cross-section. However, if information were available, it might be interesting to evaluate the effect of changes in the values of covariates over time. For example, marital status, which is usually represented by a 0-1 dummy variable, can change over the duration period as an individual becomes single, 0, or married, 1. The price of an innovation, which is typically represented by a time series, can change several times. As will be seen, this kind of explanatory variable, also called 'time varying covariates', is not a continuous function of time, but follows step-functions over time. In this regard it is different from other time variables, such as age and time itself, which change continuously as a function of time. For the latter, which are called 'time dependent' covariates, a different treatment can be given.[8]

The proportional hazard model

The hazard function can be re-formulated to allow for the influence of explanatory variables. Let X be a vector of time invariant covariates with a vector of unknown parameters β. The hazard can be expressed by

$$h(t,X,\theta,\beta) = h_0(t,\theta)\ q(X,\beta),$$

where $h_0(t,\theta)$, which is known as the 'baseline hazard', assumes any parametric distribution, such as those exemplified in Table 7.2. Models with this specification are called 'proportional hazards'. The covariates enter $q(X,\beta)$, for which the most widely used and convenient specification is

$$q(X,\beta) = \exp(\beta' X).$$

This form guarantees the necessary nonnegativity without imposing restrictions on β. Also, log-linearization allows an easy partial-derivative interpretation of parameters:

$$\frac{\partial \ln h(t,X,\theta,\beta)}{\partial X} = \frac{\partial \ln q(X,\beta)}{\partial X} = \beta.$$

Note that covariates act multiplicatively on the baseline hazard function and the signs of β are interpreted as the effect of explanatory variables on the conditional probability of completing a spell.

The partial-likelihood approach

The proportional hazard model, as presented above, has two possible deficiencies. First, the time dependence of the hazard is incorporated by electing a specific parametric distribution for the baseline hazard. In the absence of theoretical guidance for the appropriate functional form, this can be a strong ad hoc imposition. Second, variables that change over time are not incorporated. These problems can be overcome by applying the partial-likelihood approach suggested by Cox (1972, 1975) to estimate the proportional hazard parameters.

Consider n observations with ordered completed spells t_i. Suppose there is neither censoring nor tied observations. The conditional probability that the first observation leaves its initial state at duration t_1, given that any observation could have been completed its spell at t_1, is

$$\frac{h(t_1,X_1,\theta,\beta)}{\sum_{i=1}^{n} h(t_1,X_i,\theta,\beta)}.$$

Assuming the proportional hazard model, the expression reduces to

$$\frac{q(X_1,\beta)}{\sum_{i=1}^{n} q(X_i,\beta)},$$

which is the contribution of the first, and shortest, observation to the partial likelihood. The contribution of the jth observation would be

$$\frac{q(X_j,\beta)}{\sum_{i=j}^{n} q(X_i,\beta)}.$$

Note that the conditional probability implies the summation term must comprise only individuals whose spells have not been completed prior to time t_j. The ordering of spells guarantees this conditionality. The likelihood is formed as the product of the individual contributions, and the resulting log-likelihood function is

$$L(\beta) = \sum_{i=1}^{n} [\ \ln q(X_i,\beta) - \ln(\sum_{j=i}^{n} q(X_j,\beta))].$$

Censored spells are incorporated in the second term within the summation of the log-likelihood function. They do not enter the first term because no information is available on their exact duration. Ties are handled by including a contribution to likelihood for each of the tied observations, using the same denominator for each.

As can be seen, covariate coefficients are estimated without any parametric imposition on the distribution of T. Time dependent covariates enter as continuous functions of time and the data. For example, suppose X is the variable age for a set of observations. The information available is the individual age x_i expressed in units of time at $t = 0$. The values of the time variable $X(t)$ for the individual i would be $f(t,x_i) = x_i+t$. Here, time is indirectly incorporated in the partial likelihood model. Direct dependence of the hazard itself on time cannot be calculated.

Accelerated failure time model

Another way to introduce covariates in duration models is to specify the parameter λ of the baseline as a function of explanatory variables, which can be done by expressing

$$\lambda = e^{-\beta' X}.$$

The vector X includes a constant term and, for the moment, a set of time invariant variables. The exponential hazard function, for example, would be expressed exactly as above. Covariates act multiplicatively on time and not on the entire baseline, as in proportional hazard models. The baseline still exists, but the time axis is re-scaled by the covariates. In other words, the set of X variables can 'accelerate' or 'decelerate' the individual duration in a particular state.

The models presented in the last two sections did not include covariates that follow step functions, or time varying covariates. Greene (1992), based on Petersen's work (1986a, 1986b), incorporates this kind of variable in the accelerated failure model.[9] The following paragraphs summarize the basic formulations necessary to allow that inclusion.

Suppose that t_k, the end or the censored time of a spell, can be divided in k non-overlapping and adjacent segments of time, where $t_0 = 0$ and $t_0<t_1<t_2<...<t_j...<t_k$. The intervals are not necessarily equal in terms of length. Suppose z is a vector of covariates that follow step-functions over time, that is, time varying covariates; the vector value, at duration t, is z(t) and stays constant for finite sub-periods of time. In other words, during the duration interval t_{j-1} to t_j, z remains constant at $z(t_{j-1})$ and jumps to $z(t_j)$ at t_j. Let us call Z(t) the path taken by z up to duration t. From relation 7.9, the probability that a spell is of length at least t_j, given it lasts at least until t_{j-1} and given the path taken by z up duration t_j, can be expressed by

$$P[T\geq t_j \mid T\geq t_{j-1}, Z(t_j)] = \exp\left[-\int_{t_{j-1}}^{t_j} h(s \mid Z(t_{j-1}))\, ds\right]. \qquad 7.13$$

As the covariates stay constant from t_{j-1} to t_j, the integration term can be reduced to a more simple function. For example, assuming the accelerated failure time model of the last section and an exponential distribution for T, the integrated hazard in 7.13 is λt. In this case, λ stays constant during the interval, as long as the covariates do the same. However, it may change from one interval to the next. Given expression 7.13, a survival function for the entire period can be calculated allowing for covariates that follow step functions. Thus, the probability that a spell is of length at least t_k conditional on $Z(t_k)$ is

$$S(t_k|Z(t_k)) = \prod_{j=1}^{k} Prob(T\geq t_j|T\geq t_{j-1}, Z(t_j))$$

$$= \exp\left[-\int_{0}^{t_1} h(s|Z(t_0))\, ds\right] \times .. \times$$

$$\exp\left[-\int_{t_{k-1}}^{t_k} h(s|Z(t_{k-1}))\, ds\right]$$

$$= \exp\left[-\sum_{j=1}^{k} \int_{t_{j-1}}^{t_j} h(s|Z(t_{j-1}))\, ds\right]$$

With this expression, the density function is obtained from relations 7.6 and 7.7 and, finally, estimation follows maximum likelihood procedures using 7.11 or 7.12.

Heterogeneity and specification tests

Explanatory variables are introduced in econometric models in order to control for possible differences in the distribution of the dependent variable among individuals in a population. If there is complete control, any initial heterogeneity is offset and a perfect statistical model is obtained. In practical terms, however, the true set of variables is rarely known and we can control only for the observed ones. A model with an incomplete set of variables may or may not be acceptable, depending on the level of remaining heterogeneity after control. In duration analysis, heterogeneity has a special effect. When it is present, the estimate of duration dependence is biased downwards and can induce inaccurate inferences. For example, one can estimate the hazard is decreasing over time when it is actually constant. A possible treatment for this problem is carried out by inserting an error term in the hazard (Lancaster, 1979). However, if the duration dependence parameter is largely positive, and if qualitative results are more important than accuracy, an excessive parameterization may not be worthwhile. Instead, it might be better to detect misspecification by using known tests and, if necessary, review the model to reduce heterogeneity.

Hypotheses about estimated parameters are generally examined by employing the well-known tests for analysing maximum likelihood estimates: Wald, Lagrange Multiplier, and Likelihood Ratio. The choice is a matter of convenience. The usual approach is to test the joint hypothesis that all coefficients, except the constant, equal zero. It shows the importance of explanatory variables to the fit of the model.

An assessment of the overall specification of the model is possible through the examination of the integrated hazard. An important property of the (minus) logarithm of the survival function, or the integrated hazard, is the fact that it has an exponential distribution with parameter $\lambda=1$, whatever the distribution of the survival function. The proof is drawn from the following result in mathematical statistics: If f_X (x) is the probability density function of X, the density of the random variable Y=g(x) is

$$f_Y(y) = \frac{f_X[g^{-1}(y)]}{dy/dx},$$

where the numerator is the density function of X expressed in terms of y.

Applying the above result to Y = −log S(t), gives

$$f_Y(y) = \frac{f_T\,[S^{-1}(e^{-y})]}{dy/dt}. \qquad 7.14$$

Where the numerator is the density function of t expressed in terms of y.

The denominator of 7.14 can be rewritten as

$$\frac{dy}{dt} = \frac{d[-\log S(t)]}{dt} = h(t) = \frac{f_T(t)}{S(t)}.$$

Where the last fraction can be expressed in terms of y, giving

$$\frac{f_T\,[S^{-1}(e^{-y})]}{S[S^{-1}(e^{-y})]} = \frac{f_T\,[S^{-1}(e^{-y})]}{e^{-y}}. \qquad 7.15$$

Substituting 7.15 into 7.14 we obtain $f_Y(y) = e^{-y}$, which is the unit exponential density distribution.

A specification test for a fitted model can be conducted by checking whether the estimate of the integrated hazard (also known as generalized residuals) has a distribution similar to the unit exponential. In practical terms, however, most samples are right-censored. In these cases, the check would be to determine whether the integrated hazard looks like a censored sample from a unit exponential distribution. An extension of the method is to plot the integrated hazard of the generalized residuals against the generalized residuals. For a well specified model the points should lie approximately along a 45-degree line. This comes from the fact that the true values of minus the logarithm of the survival has density distribution and related survival equal to a unit exponential distribution, that is $f(y) = S(y) = e^{-y}$. Since $-\log S(y) = y$, the true points must lie exactly on a 45-degree line.[10]

Examples of the application of duration analysis to technology adoption

To suggest possible applications of duration analysis in technology adoption, three examples are briefly outlined in this section. Numerical results are not shown, given that our major interest is only to illustrate the sort of problems duration analysis can deal with and what analytical results can be achieved.

Hannan and MacDowell (1984) are referred to as the first to introduce duration analysis in technology adoption studies. Using data on the adoption of automatic teller machines by firms in the banking industry, they examined the part played by the competitive environment in which firms operate in the decision to adopt an innovation. The sample comprised 3,841 banking firms, of which 740 had adopted the new technology by the end of the 1971-1979 period. Data included annual observations of bank and market characteristics. The model was specified by employing an exponential proportional hazard function $h(t) = \exp(\beta' X_t)$, where the vector X_t incorporates explanatory variables and a constant term, which absorbs the baseline hazard. Twelve explanatory variables were tested, including dummy variables used to represent regulatory environments, such as branching prohibition or restriction. The analysis resulted in the Schumpeterian conclusion that:

> larger banks, banks operating in more concentrated local banking markets, and banks that are part of banking holding company organizations evidence a higher conditional probability of adoption of this new technology, all else equal (Hannan and MacDowell, 1984, p.334).

Later, Hannan and MacDowell (1987) extend their approach to investigate the nature of firms' reactions to rival precedence in the adoption process. Prior adoption variables were included and statistically accepted, resulting in the empirical finding that rival precedence increases the conditional probability of adoption. Other explanatory variables, such as size and a proxy for market concentration, were tested and eventually accepted. Additionally, the variable time was also introduced, giving the baseline hazard a Gompertz distribution assumption. The positive and significant value of the time dependence parameter led the authors to suggest that adoption may have been stimulated by technological improvements in the innovation over time, or by reduction in uncertainty as more information becomes available on the experience of actual adopters.

In the same study, duration analysis was also used to test the simple version of the epidemic model. According to the latter, diffusion of an innovation is represented by a logistic curve, which depicts adoption as a function of the population which has already adopted. The authors suggested this relation can be expressed by an exponential hazard function:

$$h(t,D_t,\beta) = \beta D_t = \exp(\ln\beta + \ln D_t),$$

where D_t is a time variable for the diffusion level, the proportion of banks that had introduced ATM system prior to year t, and β is the related parameter. In fact, the model is a restricted version from a broader one in which firm-specific variables are added. Hannan and MacDowell's data supported the acceptance of these variables and, consequently, the simple epidemic model was considered inadequate for explaining diffusion. Also, the inclusion of time as a variable into the hazard function, and eventual acceptance of a positive related parameter, leads to the rejection of the logistic curve in favour of an alternative functional form.[11]

The epidemic model was tested again in Karshenas and Stoneman's (1990) study on diffusion of new process technology. The authors point out that the theoretical development in the area had advanced beyond the empirical work and suggested a model to test the outcome already achieved. The first studies on diffusion focused on the epidemic model, which, setting the dependence of the adoption rate on the level of diffusion, brought about the importance of endogenous mechanisms. Recent contributions have concentrated on the explicit treatment of the adoption decision, which Karshenas and Stoneman suggested could be divided into three categories:

1 *Rank effects* Firms' different characteristics affect their returns from the use of new technology. Consequently, a distribution for the adoption time is obtained and a diffusion path can be built up. Rank effects refers to the threshold models presented in chapter 5.

2 *Stock effects* The diffusion of an innovation is followed by a reduction in the cost of production of users. Consequently, the industry price will decrease and output will increase. Then, returns from the new technology will fall and, for some potential adopters, adoption will not be profitable, unless the cost of adoption decreases over time. Assuming the latter really happens, a diffusion path can be generated from this game theoretic approach.

3 *Order effects* There are some economic advantages for the first adopters, such as obtaining prime geographic areas and employing the most skilled labour, and these have to be considered with the assumption that the cost of

acquisition falls over time. A firm has to weigh whether it waits and takes the advantage of falling costs or grabs the first adopters' benefits.

These three effects and the epidemic model were tested through the following hazard function:

$$h(t,X,\beta) = h_0(t) \exp(X'\beta).$$

Where X is a vetor of explanatory variables, representing rank, stock and order effects, while $h_0(t)$ is the baseline hazard, which absorbs epidemic effects. The acceptance of an exponential, or constant, baseline hazard would mean the rejection of the epidemic model, while a positive duration dependence would mean its validation. In the end, Karshenas and Stoneman's data reject the memoryless baseline in favour of a Weibull hazard with positive duration dependence. Rank effects were also accepted, although order and stock effects were rejected.

Another example of duration analysis applied to technology adoption is Levin et al.'s (1987) study on adoption of optical scanners by food stores. They made use of a sample of 468 firms, of which 132 had adopted, 151 were censored due to nonadoption, and 185 were censored because firms had left the market before adopting. Data comprised firm and market characteristics for 84 months starting on January 1974. The partial likelihood approach for the proportional hazard model was employed. This method does not include direct dependence of the hazard over time. However, given that the main objective was to establish the relationship between market structure and the time path of adoption, that feature was not taken to be of great concern. The variable effects could be estimated without any parametric imposition on the hazard. Along with estimates for the whole period, two additional sets of hazard functions were delineated: one for the first 48 months and another for the remaining time. This approach allowed to observe the change in the effect of explanatory variables on the adoption process. Some variables - such as those representing market concentration, per capita personal income, clerk/cashiers wage and firm size - played an important role in the early stage, but stopped or decreased their influence during the later period. In this second phase, prior adoption was found far more important in explaining the innovation diffusion than in the first. It is worth emphasizing that the single estimates, based on the entire period, were not able to generate these conclusions.

Finally, we have to stress that, in the above examples, parameter estimates were statistically assessed by both likelihood ratios and asymptotic standard errors. However, an overall goodness of fit test was only offered by Karshenas and Stoneman (1990). With heavily censored data, they tested how well the residuals fitted a unit exponential distribution with censoring. An exponential distribution for the residuals, $g(e_i) = \lambda \exp(-\lambda e_i)$, was estimated and the null hypothesis H_0: $\lambda = 1$ was tested against H_1: $\lambda \neq 1$. Given that the hypothesis that $\lambda = 1$ was not rejected, their model was validated.

Conclusion

In this chapter we have reviewed two econometric methods that have been used in recent studies on adoption and diffusion of innovations. Probit/logit models have proved to be useful tools to analyse adoption decisions. They have been employed in several empirical studies, including adoption of sustainable technologies (D'Souza et al., 1993). Duration analysis has only recently come to the fore and is a promising tool for empirical investigations in this area.

To put these research methods in context, consider again the diffusion models of chapter 5. The first empirical studies on diffusion of innovations, known as epidemic models, were based on the fitting of a logistic curve to aggregate data. The simple version of the epidemic model assumes that the population is homogeneous and takes the level of diffusion already reached (a proxy for an internal mechanism of information spread) as the only explanatory variable. The two-step epidemic model expands the number of explanatory variables, but with a limited flexibility due to degrees of freedom problems in the second stage of estimation. Both models suffer the criticism of improperly using aggregate data. A better model would control for differences between firms by using individual data.

Logit/probit methods are well-established approaches in the econometric literature. They were initially used in threshold models of consumer decision and later in the analysis of technology adoption decisions. However, the empirical analyses conducted using these techniques have dissociated adoption from diffusion. In general, the data used refer to a given point in time, resulting in a static model of adoption. The dependent variable (adoption, non-adoption) does not pick up adoption over time, as it does not allow for firms' different waiting times. Besides, explanatory variables usually refer to the point in time when the data were collected.

The greatest advantage of duration analysis over the preceeding methods is that it deals with both cross-section and time series data. Firms' characteristics, innovation price, output price, environmental characteristics and other explanatory variables may change not only from one individual to another but also over time. Duration analysis permits one to process information on these two types of change. Adoption and diffusion are investigated together. In other words, adoption can be explained as a dynamic process, while in the probit/logit models it is explained in a static manner.

Time series variables, such as output and innovation prices, which may influence the adoption decision, may not be employed in logit/probit models, because they usually do not vary from one individual to another. In duration analysis, variation in covariates over time is an alternative to variation between individuals. Thus, this type of information is not lost. Although the great advantage of duration analysis is its ability to deal with variables that change over time, it is not always feasible to obtain information on the past. This is particularly the case of variables representing individual characteristics. Sometimes, it may be necessary to assume some variables as time invariant and combine them with available time variant covariates. Moreover, the method (especially models with time varying covariates) is still a new area in econometric research. As pointed out by Kiefer (1988, p.671),

> Identification is tricky, in that the effect of trending regressors is difficult to separate from possible duration dependence. Thus, for estimates to be precise, the time paths of regressors must vary substantially across individuals. The practical problems of specifying, estimating and interpreting models with time varying explanatory variables are still important areas of active current research.

Notes

1. For details and exemplification of the different classes of models see Maddala (1983) and Greene (1993).
2. Although the logistic distribution offers mathematical convenience, the availability of fast computer programmes for estimation makes this advantage no longer important. For detailed comparison and other possible distributions, see Maddala (1983).
3. See Greene (1993).
4. The heterocedasticity test presented here is not suitable to logit models (see Greene, 1993).
5. Duration analysis has been incorporated in computer packages applied for social science, such as SAS, SPSS, and LIMDEP.
6. In technology adoption studies, when some firms leave the market before the end of the observation period, right-censoring is done at the leaving

date. Censoring at the beginning of a process is also possible. In this case, left-censoring applies, although it is less common in social science studies.

7 Non-parametric methods of duration analysis can be useful as a starting point for parametric estimation. Life tables and survival curves based on actuarial methods are derived from data on duration time. Also, graphical plots of the data can suggest functional forms for a parametric version. Another non-parametric method of analysis is the Kaplan-Meier or product limit estimator. Given our particular concern with parametric estimation, these methods will not be presented here.

8 For a broad classification of covariates see Kalbfleisch and Prentice (1980) and Lancaster (1990). The definitions presented here aim to be helpful not only in the model construction but also in providing an understanding of some of the restrictions on the use of available computer packages. Some of the latter are not able to estimate models with either time varying covariates or time dependent covariates.

9 Time varying covariates can also be incorporated in proportional hazard models. For detailed explanation, see Collet (1994), Lancaster (1990), Cox and Oaks (1984), and Kalbfleich and Prentice (1980).

10 For further discussion, including numerical tests, see Lawless (1982), Kalbfleisch and Prentice (1980), Collett (1994), Lancaster (1990), and Kiefer (1985).

11 Hannan and McDowell's argument cannot be used to reject the two-step epidemic model, in which other explanatory variables are used to explain the variance of the diffusion speed parameter (see chapter 5). Also, alternative functional forms for the logistic curve have been suggested in the literature on the epidemic model.

8 Adoption of sustainable agricultural technologies in the State of Espírito Santo

Introduction

In the early 1980s, few farmers in Espírito Santo were using alternative sustainable technologies in agriculture. In the middle of the decade, however, the number of farmers changing to LEISA practices, such as organic fertilization and use of cheap methods of plant protection started to grow. Two factors might have contributed to the increased diffusion of these techniques: decreasing coffee prices and the extension service provided by NGOs.

Many small farmers might have responded to the decline in coffee prices by reducing the use of external inputs and returning to unsustainable subsistence farming. Since rural wages had decreased, labour-intensive LEISA would have become economically advantageous. However, information about these practices was not widely diffused, as up to the middle of the 1980s, there was no institutional mechanism to do this. These technologies were viewed with suspicion and the first adopters were generally discriminated against by their neighbours.

By the end of the 1980s, several institutions - unions, farmers' associations, NGOs, the Landless Movement, the Lutheran Church, and agricultural schools - became interested in finding practical solutions to alleviate rural poverty in the region. In 1981, the FASE, a Brazilian NGO, launched the Alternative Agriculture Project (PTA) to promote adoption of alternative technologies by small farmers in Brazil (chapter 2). During the 1980s, the PTA established a national network of offices in ten Brazilian States. In 1986, the organization established the Espírito Santo branch and promoted meetings among farmers and local institutions to discuss the feasibility of alternative systems to solve farmers' problems. In the following year an effective process of information diffusion started. Since then, the Alternative Agriculture Programme Association (APTA), its new regional name, has been heading a network of institutions in the State, which are devoted to the promotion of small farmers' organizations and the diffusion of sustainable agricultural technologies (Table 8.1).

The determinants of the adoption/diffusion process are investigated in this chapter. It is hypothesised that, as well non-economic characteristics of the farm enterprise, the economic environment was important to the adoption/diffusion process. A survey was conducted in 1994 to obtain data on adopters and non-adopters.

Table 8.1
Some organizations that deal with sustainable agricultural technologies in the State of Espírito Santo

MEPS Educational and Promotional Movement of Espírito Santo. These agricultural schools have used innovative methods of education and diffusion of agricultural sustainable technologies.

Augusto Ruschi Ecological School Horticultural experimentation station in Cachoeiro de Itapemirim, which became nationally famous due to its experiments in chemical-free vegetable production.

CIER Rural Education Integrate Centre. This governmental rural school has developed and diffused sustainable practices for soil recovering in Boa Esperanca.

Mendes da Fonseca Experiment Station The station has developed and evaluated LEISA practices in the State.

Guandú Project The Project was initiated in 1987 by the IECLB (Lutheran Church) to provide assistence for small farmers' communities. It has promoted social organization and diffusion of sustainable agricultural technologies. Two farmers' associations were created directly as a result of the project: the APPRSP (Serra Pelada Small Farmers' Association) and APPRLT (Laranja da Terra Small Farmers' Association). Both of them have most of their members dealing with LEISA.

APSAD-VIDA Santa Maria Farmers' Association for Life Protection; an association of farmers who have adopted organic production methods.

APTA Alternative Technologies Programme Association. This non-governmental organization was created in 1986 to identify, develop, and diffuse alternative technologies in Espírito Santo. APTA members are other organizations, which make up the Agroecology Network. The association itself is affiliated to the national Alternative Technology Project Network (see AS-PTA in Table 2.12).

Sources: APTA (1993), and Soares and Vivan (1988)

Data and sampling method

Most data used in this study were obtained from a survey which was conducted among farmers during August and September 1994. A total of 148 personal interviews in 22 municipalities of the State yielded information on adopters and non-adopters of sustainable agricultural technologies. The sampling method could not be based on a random selection due to two reasons: (1) the number of adopters in relation to the total number of farmers in the State was expected to be small, which would make it difficult to obtain a satisfactory number of observations from a totally random selection; and (2) a complete list of adopters, from which a sample could be drawn, did not exist. Thus, adopters were selected with the help of experts from APTA. They were asked to provide the best possible representation of farmers using LEISA technologies (see Table 4.2 for definitions of these technologies) in the State. For each adopter, a nearby conventional farmer was randomly chosen.[1] After inspection of the questionnaires, seven observations were dropped due to missing data. The final sample consisted of 64 adopters and 77 non-adopters.

A farm sustainability index was built to check the sample reliability. Weights were assigned to different practices, according to how strong each one is believed to influence sustainability (Table 8.2).[2] For example, if only organic fertilizer was used to maintain soil fertility, a score of +3 has assigned. However, if only chemical fertilizers were used and use had increased in the last three years, a penalty of −2 was assigned. Use of non-chemical means for crop protection added +3, while use of chemical crop-protectors (−2) regularly (−2) implied a penalty of −4. Each farmer obtained a total score, which ranged from a minimum of −8 to a maximum of 9. Farmers whose scores were above the mean were classified as adopters, while those whose scores were equal or below the mean were considered non-adopters. In this scheme, it is possible for a farmer to use chemicals, but still be considered an adopter, if other practices that support sustainability are used. For instance, a farmer who uses a combination of chemical and organic fertilizers (though the use of chemical fertilizer has been decreasing over the last three years), and uses chemical crop-protectors (but this too has been decreasing over the last three years and is not applied regularly) has a total score of + 2, which is above the mean (+1.3).

Using this index as an indicator of sustainability, a total of 63 'adopters' and 78 'non-adopters' were identified. Comparing these two groups with those of the original sample revealed only three mismatched observations (two adopters, and one non-adopter). The empirical results presented in this chapter are based on the full set of observations and the original classification of adopters and non-adopters. Dropping the three mismatched observations and re-estimating the models did not significantly affect the results (see appendix, Tables A.1 and A.2).

Figure 8.1 shows the cumulative distribution of the number of adopters over time. The duration model, as it will bee seen in this chapter, tries to explain why some years elapsed between the first adoption in 1980 and a more widespread diffusion in the second half of the decade and the early 1990s.

Table 8.2
Farmer sustainability index

Index = $\sum W_i P_i$,   W = weights   and   P = practices

Practices/weights

Fertilization
Use of chemical and organic fertilizers: 2
Use of organic fertilizer only: 3
Use of chemical fertilizer only: −1
Chemical fertilizer use has increased in the last three years: −1
Chemical fertilizer use has decreased in the last three years: 1
Crop protection
Use of chemical crop-protectors: −2
Use of non-chemical means for crop protection: 3
Chemical crop-protectors use has increased in the last three years: −1
Chemical crop-protectors use has decreased in the last three years: 1
Chemical crop-protectors applied regularly: −2
Application of chemical crop-protectors exceeds the prescribed amount: −2
Other
Mix-crop system: 2
Soil analysis in the last three years: 1

Descriptive statistics of index values

Number of observations: 141	Mean: 1.3
Maximum value: 9	Number of observations above the mean: 63
Minimum value: −8	Number of observations below the mean: 78

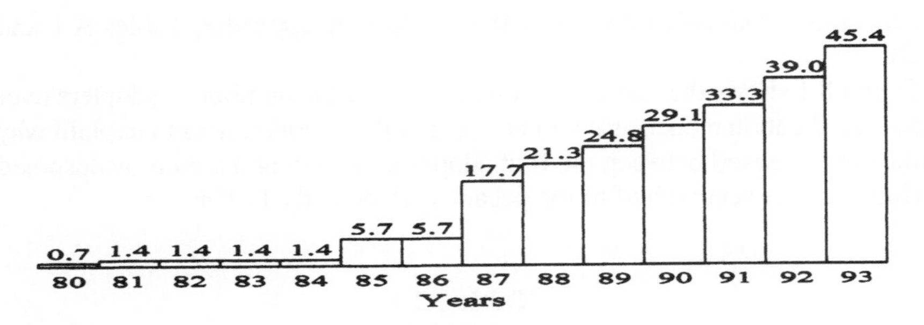

Figure 8.1 Cumulative distribution of the number of adopters in a sample of 141 farms, Espírito Santo (% of the sample)

Main farming activities

Table 8.3 shows the shares of the major groups of activities in farms' total area during the agricultural year 1993-94. These were not very different from the ones for the State as a whole in 1985 (Table 3.3). Adopters and non-adopters were broadly similar in this respect. Crops and pastures occupied three quarters of the total area, and coffee was the main crop for both groups of farmers (Table 8.4). Adopters, however, were more diversified; 87% of their crop production value derived from coffee, beans and horticultural products, which includes a variety of fruits and vegetables (mainly coconut, oranges, beetroot, potatoes, carrots, cabbage, green beans, and lettuce). Non-adopters obtain the same percentage with only three traditional products: coffee, beans, and maize.[3] After adoption of alternative practices, farmers reduced their production of tomatoes (20 farmers) and garlic (21 farmers). Both activities were considered economically difficult without chemicals. Only 3 adopters in the sample were new entrants to agriculture.

The production values used to calculate the percentages of Table 8.4 were obtained from quantities and prices reported by farmers in the sample. Coffee prices exceptionally increased after July 1994 due to frost in South Brazil (Figure 3.1). Many farmers sold their products at these higher prices, thus inflating coffee production value and income in that particular year.[4] Thus, in preceding years, the shares of coffee in the total value of crop production were smaller than those of Table 8.4. This illustrates the importance of coffee prices in determining farmers' income.[5]

Most adopters did not receive a premium price for their products. Only four of them answered positively when asked if the prices they obtained in the market were higher than those received by conventional farmers. Neither adopters nor non-adopters were supported by the PGPM (chapter 2).[6] Moreover, only 16 adopters in the sample obtained rural credit before adoption, and, after conversion, the number of contracts fell to five. In 1993, only five adopters and one non-adopter took loans from the rural credit system.

Reasons to adopt: farmers' perceptions

Adopters were asked to indicate the importance of each of potential reason for adopting the alternative agricultural practices. The scoring scheme and results of this exercise are shown in Table 8.5. Non-economic reasons - family health, empathy with the aims of the ecological agriculture, and religious considerations - obtained higher mean scores than economic reasons, although the latter could not be considered irrelevant. 'Low cost of alternative agriculture', 'less risk', 'price of chemical inputs', and 'higher profit', obtained mean scores which

Table 8.3
Land use by major farming activity (%), 141 farmers

	Perman. crops	Annual crops	Set aside	Pastures	Natural forests	Planted forests	Non-agr. land
Adopters	29.2	13.6	9.4	32.2	13.8	0.5	1.2
Non-adop.	28.2	15.1	6.1	35.1	12.2	1.2	1.8

Table 8.4
Share of the main crops in the 1994 crop production value, 141 farmers

Adopters			Non-adopters		
Crop	*%*	*Cum.%*	*Crop*	*%*	*Cum.%*
Coffee	66.6	66.6	Coffee	53.8	53.8
Horticulture	16.2	82.8	Beans	20.1	73.9
Beans	4.4	87.2	Maize	14.0	87.9
Bananas	3.6	90.8	Horticulture	7.0	94.9
Maize	2.8	93.6	Bananas	3.5	98.4
Rice	2.7	96.3	Rice	1.2	99.6
Cassava	1.8	98.1	Cassava	0.3	99.9
Other	1.9	100.0	Other	0.1	100.0
Total	100.0		Total	100.0	

Table 8.5
Farmers' perception of the main reasons for adoption*

Sample of 64 farmers and sub-sample of those who were still using chemicals (30 observations)

Reasons	*Full sample*		*Using chemicals*	
	Mean	St. dev.	Mean	St. dev.
Family health	2.8	0.5	2.8	0.5
Empathy with the aims of ecological agric.	2.2	0.8	2.2	0.9
Religious	1.5	1.1	1.6	1.2
Low cost of alternative agriculture	1.4	1.3	1.8	1.2
Less risk	1.3	1.2	1.6	1.4
Price of chemical inputs	1.2	1.2	1.6	1.1
Higher profit	1.1	1.2	1.4	1.3
Less dependence on banks	0.8	1.2	1.1	1.4
Higher prices for alternative products	0.7	1.1	1.0	1.2
Non-response of chemical fertilizers	0.6	1.0	0.7	1.0
Difficulty in getting labour	0.4	0.8	0.6	0.9
Rural credit interest rate	0.4	1.0	0.6	1.2

* Farmers were asked to score the importance of the above reasons according to the following scale: 0 - Not important, 1 - Somewhat important, 2 - Important, and 3 - Very important

indicated these factors were at least 'somewhat important'. The low scores given to 'higher prices for alternative products', 'less dependence on banks', 'rural credit interest rate', and 'difficulty in getting labour', can be explained by the fact that few adopters in the sample have received market premia over conventional prices, and/or have been assisted by the rural credit policy. Finally, farmers reveal that 'difficulties in getting labour' were not relevant to the adoption decision. Given that LEISA is labour-intensive, one can deduce they do not have problems in terms of labour availability.

A total of 30 adopters in the sample were still using chemicals. They were asked to indicate reasons for doing so. Table 8.6 shows the frequency of each stated reason and reveals that farmers were concerned with possible yield and production declines, which can be linked with their economic rationale. As shown in Table 8.5, economic considerations were more important for this category of adopters than they were for the sample as a whole.

Sources of information: farmers' perceptions

Adopters were asked to indicate the importance of several sources of information about alternative technologies (Table 8.7). Two sets of scores were given: one referring to the importance of the sources to the farmer's decisions at the time of adoption (columns 1 and 2); and the other referring to the importance of the sources to the farmer's decisions in 1994 (columns 3 and 4). The mean scores can be used to rank the sources in order of importance and to detect changes over time. In column 1, the highest scores were obtained by 'other', 'farmer associations', and 'religious institutions'. In the 'other' category, the names of non-governmental organizations (other than farmers' associations and religious institutions) were mentioned 30 times by farmers (from a total of 34 farmers who opted for this category). Comparing the mean scores of column 1 with those of column 3, one can see that the importance of farmers' associations has increased, while that of 'other' sources (NGOs) has decreased. The fact that these NGOs were more important at the time of adoption than after may reflect their strategy, which was to promote LEISA and stimulate farmers' organization.

Table 8.6
Reasons for still using chemicals and corresponding number of adopters who indicated them

Sample of 30 adopters

Reason	Number
Insects could get out of control	7
Weed could get out of control	1
Crops could be destroyed by diseases	7
Yields could decrease	11
Labour is difficult to find	4
Profit would be impossible without them	3
Other (non-specified)	6
Other (specified by farmers)	
Organic fertilization on coffee is difficult	3
Chemicals are used to control livestock parasites	2
As sharecropper, farmer had to use chemicals	1
On-farm trail to compare different practices	1
Chemicals are important to current soil condition	1

Table 8.7
Information sources on alternative agriculture and their degree of importance, 64 adopters*

Sources	*At the time of adoption*		*Currently (1994)*	
	Mean (1)	St. dev. (2)	Mean (3)	St. dev. (4)
Radio	0.5	1.0	0.6	1.1
Television	0.9	0.9	1.4	1.1
Specialized magazines	0.6	0.8	0.7	0.9
Leaflets	1.1	1.0	1.6	1.1
Other farmers	1.0	1.0	1.3	1.2
Farmers' associations	1.2	1.3	2.0	1.3
Governmental extension	0.4	0.9	1.3	1.1
Religious institutions	1.2	1.1	1.5	1.2
Other	1.4	1.4	1.1	1.2

* As Table 8.5

Table 8.8
Information sources on agriculture in general and alternative agriculture, and their degree of importance, 77 non-adopters*

Sources	*Agriculture*		*Alternative*	
	Mean	St. Dev.	Mean	St. Dev.
Radio	1.1	1.0	0.9	1.0
Television	2.1	0.8	1.9	1.0
Specialized magazines	1.4	1.2	1.4	1.2
Leaflets	1.6	1.1	1.6	1.2
Other farmers	1.7	1.0	1.9	1.1
Farmer associations	1.7	1.2	1.5	1.2
Governmental extension	1.8	1.3	1.5	1.2
Religious institutions	1.1	1.2	1.1	1.2
Other	0.5	1.1	0.4	1.0

* As Table 8.5

Table 8.9
Reasons for non-adoption of alternative technologies and number of non-adopters who cited them, 72 non-adopters*

Reason	Number
I do not have sufficient knowledge about these practices	48
I do not believe they are efficient methods to control of pests and diseases and to provide adequate fertilization	7
I am unsure about the efficiency of these methods to control pests and diseases and to provide adequate fertilization	25
They are not economically feasible (decrease profit)	14
I do not have enough time to use these practices	31
Management is difficult	32
Labour is hard to find	28
My farm does not have problems with pests, diseases and soil fertility	12
Other	5

* Farmers who knew or had heard of alternative agricultural techniques were asked to indicate, from a list of nine, the three most important reasons for non-adoption.

In addition, the importance of the local government extension service in providing information about alternative agriculture has increased. Some local offices have recognized the advantages of LEISA in alleviating farmers' economic problems and reducing environmental degradation (Espírito Santo, 1991). The mass media, especially television, have also given more attention to sustainable technologies.

Non-adopters were asked to indicate the importance of sources of information about agriculture in general and alternative agriculture in particular (Table 8.8). The highest mean scores were associated with 'television'. These were higher than those obtained by the governmental extension, which may reveal the effect of the economic crisis on the public services and/or the mass media effort to reach the rural population.

An important source of information about alternative technologies, according to non-adopters, was 'other farmers'. This could be taken as an indication of epidemic diffusion of information (discussed in chapter 5). However, its effectiveness in terms of adoption is doubtful, given that 'other farmers' obtained high mean scores from non-adopters (Table 8.8) but low mean values from adopters (Table 8.7).

Non-adopters who knew about alternative practices (72 in the sample) were asked to indicate three out of nine reasons for non-adoption (Table 8.9). 'Insufficient knowledge' was the predominant choice. Some farmers perceived that management of alternative practices is difficult and would require more of their time than they would wish, or would be able, to allocate. The number of those citing uncertainty about the efficiency of alternative methods was not as large as might have been expected. As already noted, non-adopters considered important the information they obtained from other farmers. Given the sample design, in which non-adopters were selected in close vicinity to adopters, they may have observed their neighbours using sustainable technologies. Successful alternative farmers might have reduced the level of uncertainty of conventional neighbours.

Assessing the determinants of adoption through logit and probit models and duration analysis

The adoption process is examined in this section through the econometric techniques presented in chapter 7: logit/probit models and duration analysis. A number of explanatory variables (see definitions in Table 8.10) were tested as potential determinants of adoption. Table 8.11 shows descriptive statistics of the time invariant variables, and Figure 8.2 depicts time paths of time varying covariates. The explanatory variables correlation matrix is presented in appendix, Table A.3.

There appears to be insufficient variation in the data with respect to the age, education and residence characteristics of the farmers in the sample to discriminate adopters from non-adopters. Inspection of the descriptive statistics shows that most farmers were not old, but their mean age suggests they might have experience in agriculture and can still be open to accept new ideas.[7] The average number of years of formal education corresponds to primary schooling, but indicates at least basic mathematics and literacy.[8] Middle age and low level education contradict previous findings that younger and better educated farmers are more likely to adopt (chapter 6).

Most farmers in the sample live on the farm, are landowners, and rely essentially on on-farm income to survive. These characteristics match the findings of studies about adopters of agricultural sustainable technologies (chapter 6). However, as showed in Table 8.11, the variables RESIDENCE, OWNERSHIP, and OFF-INCOME cannot clearly distinguish adopters from non-adopters here.

Table 8.10
Definitions of explanatory variables

SIZE Farm size measured by the hectares of land, including farmland outside the main area.

ACCIDENT Dummy variable indicating farmer's perception of the negative effect of chemicals on health and environment. It assumes a value of 1 if the farmer knew of any accident caused by chemicals on the farm or in the region (before adoption, if the farmer was an adopter), and 0 otherwise.

SOCIAL Dummy variable indicating farmer's social integration. It assumes a value of 1 if the farmer frequently attends meetings of any kind of farmers' organization (cooperatives, rural unions, farmers' associations).

ENVIRON Dummy variable indicating the effect of farm physical characteristics. It assumes a value of 1 if more than 50% of the farm land has flat/undulating topography and there is a stream as a water source, and 0 otherwise. These are characteristics of fertile lands at the bottom of hills (*varzeas*).

F-LABOUR Number of family members working on the farm, including the farmer.

EXT-NGO Dummy variable assuming a value of 1 if the farmer has contact with non-governmental extension service, and 0 otherwise.

NGO_t Dummy time variable indicating the period of operation of the APTA (Table 8.1). It assumes a value of 0 up to 1985, and 1 thereafter.

$R\text{-}TRADE_t$ Time variable indicating the evolution of the annual rate of change in the terms of trade for the agriculture of Espírito Santo. Terms of trade in a year t, T_t, are the IPR (index of prices received by farmers) divided by the IPP (index of prices paid by farmers) in that year. $R\text{-}TRADE_t = (T_t - T_{t-1})/T_{t-1}$.

$WAGE\text{-}CHE_t$ Time variable indicating the evolution of the rural wage in relation to the prices of chemical fertilizers and crop protectors. It is the Fundação Getúlio Vargas' index for seasonal-labour rural wage divided by an index of chemical fertilizers and crop protectors prices (average price of 28 types of fertilizers and formulations, and 101 types of crop protectors). 1980 = 1.

EXT-GO Dummy variable assuming a value of 1 if the farmer has contact with the governmental extension service, and 0 otherwise.

AGE Farmer's age.

RESIDENCE Dummy variable assuming a value of 1 if the farmer lives on the farm, and 0 otherwise.

EDUCATION Years of schooling.

OWNERSHIP Dummy variable assuming a value of 1 if the farmer owns the farm, and 0 otherwise.

OFF-INCOME Percentage of off-farm income in the farmer's total income.

Table 8.11
Descriptive statistics of the time invariant variables in the adoption model, sample of 141 farmers

Variable Name	*Adopters* Mean	Stand. dev.	*Non-adopters* Mean	Stand. dev.
SIZE (ha)	24.2	29.6	47.9	67.9
ACCIDENT (0,1)	0.7	0.4	0.4	0.5
SOCIAL (0,1)	0.8	0.4	0.4	0.5
ENVIRON (0,1)	0.3	0.5	0.1	0.3
F-LABOUR	4.4	3.1	3.5	2.7
EXT-NGO (0,1)	0.7	0.5	0.2	0.4
EXT-GO (0,1)	0.2	0.4	0.4	0.5
AGE	43.5	13.7	45.5	12.8
RESIDENCE (0,1)	0.9	0.3	0.8	0.4
EDUCATION (years)	8.1	2.7	8.1	3.2
OWNERSHIP (0,1)	0.8	0.4	0.8	0.4
OFF-INCOME (%)	17.8	27.8	20.5	32.2
Number of observations	64		77	

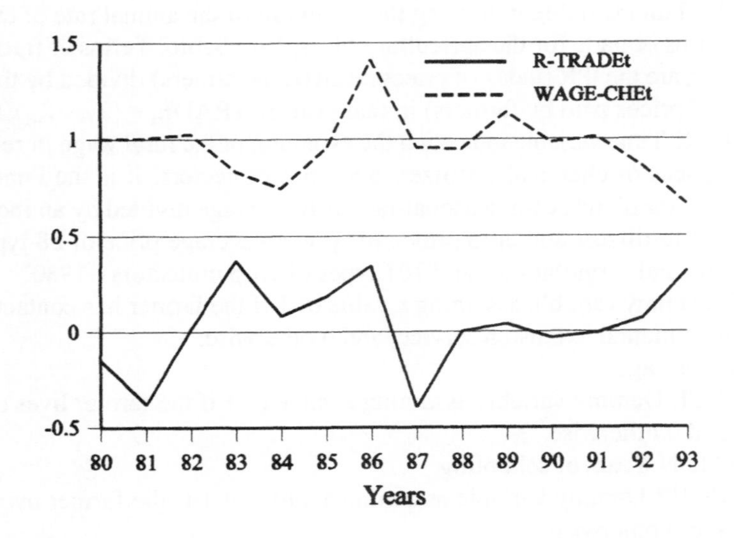

Figure 8.2 Values of R-TRADE$_t$ and WAGE-CHE$_t$

Logit and probit models

The results from fitting logit and probit models are presented in Table 8.12. Equations 1 and 3 comprise the full set of variables, while equations 2 and 4 remove the least significant ones. The likelihood ratio (LR) was used to test the hypothesis that all the slope coefficients in the probit and logit models are zero. The restricted log likelihood value is -97.134. The unrestricted log likelihoods values for equations 1 and 3 are -56.319 and -57.262, respectively. The LR test statistics are therefore 81.630 for the logit model, and 79.744 for the probit model. With 12 degrees of freedom, the critical value at the 5% significance level is 21.03, and so the joint hypothesis that the coefficients on the full set of variables are all zero is rejected in both models.[9]

The marginal effects of SIZE, ACCIDENT, SOCIAL, ENVIRON, F-LABOUR, and EXT-NGO were statistically significant at at least the 5% level. EXT-GO, AGE, RESIDENCE, EDUCATION, OWNERSHIP, and OFF-INCOME were statistically insignificant at 10% level. The LR test was used to test the hypothesis that the coefficients of these six variables are zero. The LR test statistics are 9.216 for the logit model, and 9.430 for the probit model. With 6 degrees of freedom, the critical value at the 5% significance level is 12.59, and so we fail to reject the joint hypothesis that the coefficients of these variables are zero.

The negative sign of the variable EXT-GO should not lead to the conclusion that contacts with the governmental extension service were detrimental to adoption. As noted in the last section, adopters have considered the non-governmental extension service at least 'somewhat important' as a source of information about alternative agriculture. The sign of EXT-GO only reveals that conventional agriculture is the main target of the governmental extension service.

The sign of the marginal effect of ACCIDENT confirms that adopters are relatively more concerned with health and environment. Also, the effect of ENVIRON indicates that farms' physical characteristics are relevant to explain adoption; future research should attempt to incorporate more comprehensive and accurate data on this matter.

The marginal effects of SIZE and F-LABOUR shows that adopters tend to have smaller farms and rely more on family labour than non-adopters. This helps to explain why they indicated no difficulties in finding labour (Table 8.6).

The positive and highly significant coefficients of SOCIAL and EXT-NGO indicate that these less conventional means of information diffusion played a fundamental role. Adopters have been closely integrated in farmers' organisations, the formation of which has been stimulated by other non-governmental organizations. Through APTA's network, they could share experiences and receive technical assistance.

Table 8.12

Logit and probit models: marginal effects on the probability of adoption of sustainable agricultural technologies, sample of 141 farms*

	Logit				Probit			
	Equation 1		*Equation 2*		*Equation 3*		*Equation 4*	
	Estimate	Prob |t|≥X	Estimate	Prob |t|≥X	Estimate	Prob |t|≥X	Estimate	Prob |t|≥X
CONSTANT	-1.521	0.001	-0.779	0.000	-1.388	0.000	-0.696	0.000
SIZE	-0.006	0.000	-0.005	0.001	-0.005	0.000	-0.004	0.002
ACCIDENT	0.381	0.001	0.324	0.004	0.341	0.002	0.298	0.004
SOCIAL	0.291	0.017	0.282	0.017	0.258	0.022	0.248	0.020
ENVIRON	0.270	0.040	0.263	0.045	0.251	0.043	0.245	0.042
F-LABOUR	0.076	0.006	0.069	0.002	0.060	0.011	0.060	0.002
EXT-NGO	0.486	0.000	0.447	0.000	0.452	0.000	0.409	0.000
EXT-GO	-0.178	0.159	-	-	-0.192	0.102	-	-
AGE	0.006	0.316	-	-	0.007	0.192	-	-
RESIDENCE	0.019	0.922	-	-	0.036	0.834	-	-
EDUCATION	0.033	0.143	-	-	0.030	0.148	-	-
OWNERSHIP	0.286	0.099	-	-	0.213	0.141	-	-
OFF-INCOME	-0.001	0.604	-	-	-0.001	0.539	-	-
Log-Likelihood	-56.319		-60.927		-57.262		-61.977	
Restr.(slopes=0)Log-L	-97.134		-97.134		-97.134		-97.134	
Correct predictions, adopters	85.9%		82.8%		85.9%		82.8%	
Correct predictions, n-adopt.	84.4%		88.3%		84.4%		88.3%	

* Marginal effects were calculated at mean values of regressors

Table 8.13
Sustainability index, predicted y, values of independent variables, and probability of adoption of some sample observations

Farmer	Sust. index	Obser-ved Y	Predicted Y*		Values of independent variables						Prob. of adoption*	
			Logit	*Probit*	*size*	*accident*	*social*	*environ*	*f-labour*	*ext-ngo*	*Logit*	*Probit*
A	+9	1	1	1	28.0	1	1	0	11	1	0.98	0.98
B	+2	1	1	1	1.5	0	1	1	4	1	0.88	0.87
C	+1	0	0	0	480.0	0	0	0	6	0	0.00	0.00
D	-8	0	0	0	85.0	0	1	0	1	1	0.16	0.20

* The probability of adoption and predicted y were calculated according to formulas 7.3 and 7.4. The models correspond to those of equations 2 and 4, Table 8.12. The coefficients of SIZE, ACCIDENT, SOCIAL, ENVIRON, F-LABOUR, EXT-NGO, and the constant are -0.02, 1.35, 1.17, 1.09, 0.29, 1.86, and -3.24, for the logit model, and -0.01, 0.76, 0.63, 0.62, 0.15, 1.04, and -1.77, for the probit model, respectively.

Table 8.13 shows the probabilities of adoption for four representative farmers. They were chosen from the sample according to their sustainability index value (maximum and minimum values obtained for each category) and differences in explanatory variables. Farmer A, with the highest sustainability index value, obtained the highest probability of adoption. He is followed by farmer B, who presents a lower sustainability index value, and a lower probability of adoption. Farmer D, with the lowest sustainability index value, was correctly predicted as a non-adopter. Note that, although farmer C had a higher sustainability index value than the one of farmer D, the calculated adoption probability for C is lower than for D. This is explained by the fact that farmer D is socially integrated, and has been in contact with non-governmental extension service, which increases the chances of becoming an adopter; furthermore, the large farm size of farmer C reduces his probability of adoption.

Duration analysis

Information on time of adoption and other time varying covariates could not be used in the logit and probit models but was incorporated in the duration analysis. The study assumes a relationship between the conditional probability of adoption and explanatory variables. This relationship is of the form:

$$h_i(t) = \exp\ (X_t^{\prime}\ \beta),$$

where h(t), referred to as the hazard rate, denotes the probability that farmer i will adopt sustainable technologies, given he or she has not adopted before. The vector X includes the constant term and explanatory variables. This hazard function is 'memoryless', that is, it does not change as a function of time. However, it does change when time varying covariates change over time. The estimation procedure follows the accelerated failure model presented in chapter 7 and assumes an exponential distribution for the baseline hazard. The reason for choosing this specification is that the time paths of the time varying covariates used here - NGO_t , $R\text{-}TRADE_t$, and $WAGE\text{-}CHE_t$ - do not vary across individuals, which makes it difficult to separate the effect of their trend from possible duration dependence (Kiefer, 1988).

The results from fitting the model are presented in Table 8.14. Three equations were estimated: equation 1 comprises the full set of variables, equation 2 removes the most insignificant, and equation 3 removes time-varying covariates. Log-linearization allows an easy partial-derivative interpretation of parameters, whose signs indicate the direction of the effect of variables on the conditional probability of adoption.

Table 8.14, equation 1 contains all the variables included in the logit and probit models, with the time-varying covariates: NGO_t , $R\text{-}TRADE_t$, and $WAGE\text{-}CHE_t$. The former set of variables have the same signs and a similar level of significance as they had in the logit/probit analysis. The likelihood ratio was used to test the hypothesis that the coefficients of EXT-GO, AGE, RESIDENCE, EDUCATION, OWNERSHIP, and OFF-INCOME are zero. This hypothesis was accepted at 5% significance level. The same procedure was used to test the coefficients of NGO_t , $R\text{-}TRADE_t$, and $WAGE\text{-}CHE_t$. The joint hypothesis that they are zero was rejected. The inclusion of these variables lends greater explanatory power to the analysis.

It is hypothesized that periods of contraction in agricultural profit aggravate farmers' financial constraints and increase the probability of LEISA adoption; in periods of increasing profits, this probability is reduced. The rate of change in

Table 8.14
Estimates of exponential hazard functions, adoption of agricultural sustainable technologies, 141 farms

	Equation (1)		Equation (2)		Equation (3)	
	Estimate	*Prob|t| ≥ X*	*Estimate*	*Prob|t| ≥X*	*Estimate*	*Prob /t / ≥X*
CONSTANT	-4.674	0.005	-3.040	0.029	-4.904	0.000
SIZE	-0.012	0.024	-0.010	0.034	-0.009	0.062
ACCIDENT	0.878	0.009	0.801	0.016	0.699	0.021
SOCIAL	0.690	0.050	0.781	0.020	0.652	0.039
ENVIRON	0.579	0.077	0.529	0.082	0.441	0.116
F-LABOUR	0.108	0.062	0.130	0.007	0.111	0.016
EXT-NGO	0.858	0.034	0.815	0.021	0.759	0.019
NGO_t	2.280	0.000	2.253	0.000	-	-
$R\text{-}TRADE_t$	-1.055	0.140	-1.111	0.104	-	-
$WAGE\text{-}CHE_t$	-3.930	0.009	-3.914	0.005	-	-
EXT-GO	-0.220	0.564	-	-	-	-
AGE	0.014	0.309	-	-	-	-
RESIDENCE	0.446	0.445	-	-	-	-
EDUCATION	0.064	0.242	-	-	-	-
OWNERSHIP	0.337	0.420	-	-	-	-
OFF-INCOME	-0.001	0.804	-	-	-	-
Log-Likelihood	-214.987		-217.725		-252.413	

the terms of trade, R-TRADE$_t$, is used here as a proxy for the rate of change in agricultural profitability. The index is formed by the Fundação Getúlio Vargas' IPR (index of prices received by farmers) and IPP(index of prices paid by farmers) (see Table 8.10 for the precise definition). The IPR represents the evolution of farmers' agricultural revenue, while the IPP reflects agricultural cost.[10] The coefficient of R-TRADE$_t$ is significant at 10% level in equation 2, and its negative sign is consistent with the hypothesised relationship between farm profitability and adoption.

The negative sign and significance of WAGE-CHE$_t$ indicate that the conditional probability of adoption increases when rural wages become depressed relative to the price of chemicals. This result was expected, given that LEISA is labour-intensive. Considering the sign and significance of the coefficient of F-LABOUR, one can deduce that adopters reduced costs by adopting low-external-input technologies and using family labour. Household members preferred to stay on the farm rather than to migrate or take low-paid off-farm job offers.

Finally, the positive and highly significant coefficient of NGO$_t$ shows that the APTA, and the other non-governmental organizations that make up the Agroecology Network (Table 8.1), had an important role in the process of information diffusion.

The hazard rate and the adoption probability of farmers A and D are presented in Table 8.15. Farmer A adopted sustainable practices in 1992, when the calculated adoption probabilities were 92% and 88% for equations 2 and 3, respectively. Farmer D was a non-adopter and presented low probabilities of adoption. The hazard rate in equation 3 remains constant over time due to the exponential specification of the baseline hazard. However, when time-varying covariates are included (equation 2), the hazard changes over time and accelerates or decelerates the process of diffusion.[11] The probability of adoption and the hazard increase substantially after 1986. In Figures 8.3 and 8.4, their values were calculated at the sample mean values of time invariant covariates, allowing for changes in the time variant covariates of equation 2.

A goodness-of-fit check was made by considering how well the residuals fitted a unit exponential distribution with censoring. The frequency distribution of the residuals [$g(e_i)$] of equation 1, equation 2, and equation 3 were fitted with an exponential distribution, $g(e_i) = \gamma \exp(-\gamma e_i)$, and the null hypothesis $H_0: \gamma = 1$ was tested against $H_1: \gamma \neq 1$. The maximum likelihood estimates of γ are shown in Table 8.16. They confirm that equation 1 and 2 fit the data satisfactorily, but reject the model of equation 3. This reassures us of the importance of the time varying covariates explanatory power.

Table 8.15
Hazard rate and adoption probability for two representative farmers*

	Farmer A				Farmer D			
	Equation 2		*Equation 3*		*Equation 2*		*Equation 3*	
Years	Hazard	Adop. probab.	Hazard	Adop. probab.	Hazard	Adop. probab.	Hazard	Adop. probab.
1980	0.04	0.04	0.16	0.15	0.00	0.00	0.02	0.02
1981	0.05	0.08	0.16	0.28	0.00	0.01	0.02	0.03
1982	0.03	0.11	0.16	0.38	0.00	0.01	0.02	0.05
1983	0.04	0.15	0.16	0.48	0.00	0.01	0.02	0.06
1984	0.08	0.21	0.16	0.55	0.01	0.02	0.02	0.08
1985	0.03	0.24	0.16	0.62	0.00	0.02	0.02	0.09
1986	0.04	0.27	0.16	0.68	0.00	0.02	0.02	0.11
1987	0.57	0.59	0.16	0.73	0.04	0.06	0.02	0.12
1988	0.38	0.72	0.16	0.77	0.03	0.08	0.02	0.13
1989	0.17	0.76	0.16	0.80	0.01	0.09	0.02	0.15
1990	0.34	0.83	0.16	0.83	0.02	0.11	0.02	0.16
1991	0.30	0.87	0.16	0.86	0.02	0.13	0.02	0.18
1992	0.49	0.92	0.16	0.88	0.03	0.16	0.02	0.19
1993	-	-	-	-	0.05	0.21	0.02	0.20

* The hazard rates were calculated according to the expression $h_i(t) = \exp(X_i'\beta)$. The probability of adoption by the end of t years may be expressed as $1 - \exp(-\sum_{t=1}^{t}\exp(X_t'\beta))$. The coefficients of equations 2 and 3, Table 8.14, were used in the calculation. Values of the time-invariant explanatory variables are in Table 8.13.

Table 8.16
Estimated values of γ for the exponential distribution of the residuals, $g(e_i) = \gamma \exp(-\gamma e_i)$

	γ	confidence interval at 5%
Equation 1	1.038	0.977 to 1.098
Equation 2	1.037	0.975 to 1.099
Equation 3	1.527	1.376 to 1.678

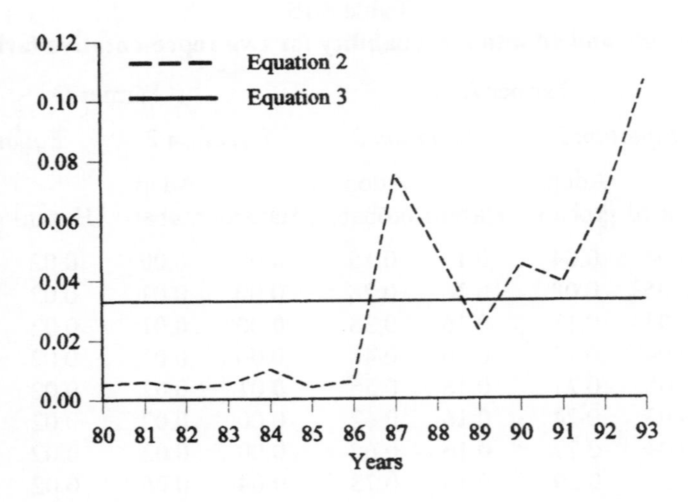

Figure 8.3 Hazard rates at mean values of time invariant covariates

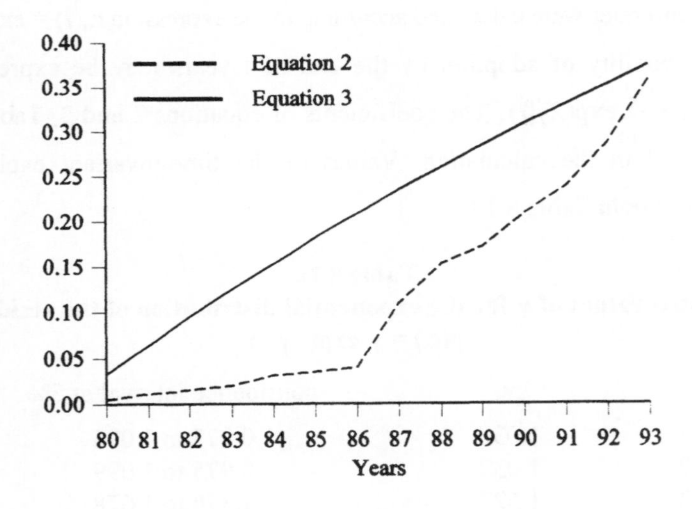

Figure 8.4 Adoption probabilities at mean values of time invariant covariates

The model of equation 2 could be improved if information on individual profitability, input and output prices were available. Thus, the use of proxy variables ($R\text{-}TRADE_t$ and $WAGE\text{-}CHE_t$), which do not change between individuals, could be avoided, and variation in the regressors' time paths between individuals could be obtained. This would allow us to increase the precision of our estimated coefficients, and to test other functional forms for the baseline hazard. This limitation, however, cannot detract from the fact that the explanatory power of equation 2 is greater than that of equation 3. Duration analysis has allowed us to highlight the effect of variables that change over time and are external to farmers.

Conclusion

It is argued that not only non-economic reasons but also economic considerations have led farmers to decide in favour of new approaches to agricultural production. Many of these factors, such as change in relative prices and government policies, are beyond the farmer's control. In Espírito Santo, the work of diffusing sustainable agricultural practices was eased by farmers' economic constraints. On the one hand, the decline in output prices squeezed agricultural profit and many farmers faced difficulties in buying external inputs. On the other hand, labour became cheap due to the economic crisis, which made low-external-input practices an attractive option for family small-holdings.

Notes

1. This sampling technique decreased the interviewers' transport costs, but also reduced the scope for discriminating adopters from non-adopters in terms of their environmental characteristics (soil types, terrain, topography, water supply, and climate).
2. Assigning weights to agricultural practices on an index of sustainability can be arbitrary, given that most technologies are site-specific. The weights showed in Table 8.2 have similarities with those of Taylor et al. (1993), who consulted a panel of experts to establish a sustainability index for cabbage farmers in Malaysia. Ideally, socioeconomic variables should be included in an index of sustainability but often, as is the case here, data are not available.
3. Data on livestock production were not available.

4 Coffee prices in the sample ranged from R$ 20 to R$ 160. From April to July, inflation was 257.6% (General Index Price, FGV), while coffee prices increased by 449.4% (Index of Prices Received by Farmers, FGV)

5 Adopters were asked whether their income before adoption was equal to, less than, or greater than that in 1994. The answers were 10, 24, and 9, respectively (18 missing observations). This outcome, however, could not be taken as an indicator of economic reasons affecting adoption, given the exceptional prices of coffee in 1994.

6 Coffee, the main crop, has not been covered by this government policy.

7 At the time of adoption, adopters mean age was 37.5 years (standard deviation of 13.7 years).

8 Formal education in Brazil comprises eight-years primary school, three-years secondary school, and four-years of undergraduate study. Five adopters and nine non-adopters in the sample were illiterate.

9 The probit model was tested for heteroscedasticity according to the specification test suggested in Greene (Greene, 1992; and chapter 7). The variance of the error term was taken as a function of several sub-sets of variables. None of the parameters in these subsets was found statistically significant, indicating that there is no heterocedasticity.

10 The IPP comprises the following items of expenditure: seeds (13.97%), fertilizers (28.43%), chemical crop protectors (10.95%), services (14.75%), fuel (14.40%), and labour (17.50%) (Monteiro, 1994). The main activities in the IPR are: coffee (45.46%), milk (13.51%), cattle-beef (8.28%), beans (4.17%), maize (4.00%), chicken (3.96%), cocoa (3.66%), and bananas (3.62%). Agricultural profitability in Brazil has been affected by inflation and government policies, mainly the PGPM and the rural credit system (chapter 2). R-TRADE_t can reflect the effect of inflation and the PGPM, although it does not allow for the direct effects of rural credit subsidies, which have been important in determining profitability of privileged farmers (chapter 2). This drawback, however, is not serious, given that few farmers in the sample have been assisted by the rural credit policy.

11 In equation 2, a major jump in the hazard is observed in 1987, which is consistent with the cumulative distribution of the number of adopters (Figure 8.1).

9 Summary and conclusions

The aim of this study was to analyse the determinants of adoption and diffusion of sustainable agricultural technologies. Several issues related to this theme were examined in order to delineate concepts and methodological approaches. Economic, social and environmental aspects of the Green Revolution in Brazil and particularly in the State of Espírito Santo were also examined as a means of establishing the background for an empirical investigation. A field survey conducted in 1994 provided the relevant data about farmers in Espírito Santo which were then analysed using descriptive statistics and two econometric approaches.

Background

Chapter 2 describes how the Green Revolution in Brazil failed to meet important criteria for sustainable development. Although aggregate production and income increased, depletion of the environment and income inequalities among farmers and between regions prevailed. These problems were exacerbated by the diffusion of high-external-input technologies and the persistence of low-external-input and unsustainable agriculture. Both governmental and non-governmental organizations have taken positive steps to remedy these problems, but these are still not sufficient to provide complete solutions.

The side-effects of the Green Revolution in the State of Espírito Santo (considered in chapter 3) are in general similar to those in Brazil as a whole. Nevertheless, some specific characteristics of the State's agrarian structure and environment defined the particular speed, the intensity and the effects of diffusion of conventional technologies. The 1960s coffee price crisis delayed the widespread introduction of high-external-input practices in Espírito Santo. Only in the second half of the 1970s when coffee prices increased, did diffusion of chemical and mechanical technologies speed up. This process decelerated in the 1980s, particularly in the second half of that decade, when government subsidies were reduced and output prices decreased. In this period, the number of farmers adopting more sustainable agricultural practices increased. The reasons for this were examined in chapter 8.

Chapters 4 to 7 provided the theoretical and methodological background for the empirical investigation presented in chapter 8. A review of several issues related to sustainable development was given in chapter 4. There it is argued that there are complex links between economic growth, environment, quality of life, concern for future generations, income distribution and poverty. This broad range of interrelated problems makes it difficult for a precise definition of sustainable development to be arrived at. The most widely accepted definition, that of the WCED, tries to balance environmental conservation and aspiration for the creation of wealth. As there is large diversity of social, economic, and environmental contexts, this definition allows for different interpretations. It assumes for instance that, in order to reach sustainable development, the developed countries should pursue policies related to conservation and rehabilitation of landscapes, while developing countries should seek equity, fairness, respect for the law, redistribution of wealth and wealth creation (Sandbrook, 1992). Uncertainty about the future (technologies and society's values) is another factor precluding precise sustainability criteria. Sustainable development is a process which induces societies constantly to re-access their goals. In agriculture there are disagreements about which technologies are actually sustainable and which are not. For instance, adopters of organic farming would consider application of any amount of chemical pesticides as unsustainable, while LEISA and IPM adopters accept chemicals as long as they are kept to certain limits. There is a consensus however, in that both conservation and better standards of living must be pursued simultaneously in the long-run. Sustainable farmers are not necessarily those who adopt rigid sets of practices regulated by governmental bureaus or by non-governmental organizations, but rather those who intentionally make a move towards this consensual goal.

Chapter 5 showed that there has been continuous progress in the theory of adoption and diffusion of technologies since the first application of the epidemic model. Recent studies have been more concerned with modelling individual decision-making behaviour than with exclusive examination of aggregate diffusion. As diffusion is a result of individual decisions over time, both aspects of the problem can in fact be analysed together. The threshold models are examples of how a diffusion process can be obtained from individual decision analysis.

Chapter 6 provided a review of several factors which may promote the adoption of sustainable technologies. It appeared that both economic and non-economic reasons have led farmers to decide in favour of new approaches to agricultural production. It was explained that many factors affecting profitability (e.g. government policies and market prices) are beyond farmers' control and can positively or negatively influence the decision to adopt sustainable practices. Our review showed that in several countries there is a shift towards integration of agricultural and environmental policies. More information and also developed markets are now available and later adopters have been better accepted socially. Some sustainable practices are becoming economically feasible and, as economic and social barriers are reduced, new entrants come to receive more positive signs than those encountered by previous adopters.

Most empirical studies of technology adoption rely on econometric models to investigate the effect of observed variables on firms' or farmers' decisions to adopt innovations. Logit/probit models (chapter 7) are well-established approaches and have been employed in several empirical studies of technology adoption and also in studies of adoption of sustainable technologies in agriculture. However, empirical analyses conducted with application of logit/probit techniques have dissociated adoption from diffusion. It is rarely considered that firms/farmers adopt an innovation at different points in time. Available data generally refer to the time of the cross-section survey instead of the time of adoption. Moreover, the effect of variables such as price, which may not change from one individual to another but according to the passage of time, cannot be investigated.

Duration analysis (chapter 7) has an advantage over the preceding methods in the sense that it uses both cross-section and time series data, thus permitting an investigation not only of the time of adoption, but also the effect of explanatory variables that change over time. Thus, adoption can be explained as a dynamic process. However, panel data on a set of farms observed over time are rarely available. One solution is to combine cross-section data on individual characteristics with time series data, such as output and input prices, which are easier to obtain. This allows an assessment of the influence of time-varying variables that are beyond the farmers' control but that can affect the adoption decision.

Main results

Logit/probit models and duration analysis were used to examine the effect of several variables on the decision to adopt sustainable agricultural practices in the State of Espírito Santo. From the logit/probit analysis it was concluded that the probability of a farmer's adopting sustainable technology increased if he/she was more integrated with farmers' organizations, had contact with non-governmental organizations, was aware of the negative effect of chemicals on health and the environment, could rely on family labour and had his/her farm located in an area of better soil conditions. On the other hand, the probability of adoption was reduced by increases in farm size. These results were corroborated by the duration analysis, but, in addition, the impact of time-varying, economic variables outside the farmers' control was examined. Here it was evident that changes in relative prices could influence the rate of diffusion. Specifically, the diffusion of sustainable technology accelerated when the decline in output prices squeezed agricultural profit and many farmers faced difficulties in buying external inputs. On the other hand, when labour became relatively cheap due to the economic crisis, low-external-input practices became a more attractive option for family smallholdings.

The above suggests that any increase in output prices and rural wages relative to the prices of external inputs leads to a decrease in the speed of diffusion of sustainable agricultural technologies. In Espírito Santo, this can happen when coffee prices increase markedly. In such periods some adopters can even reverse their decisions and increase the use of external inputs, including agrichemicals (although our data and modelling framework do not permit to test this hypothesis). To counter this and to promote the diffusion of LEISA practices, one could suggest policy interventions that penalize high-external-input agriculture. However, their short and long-term effects on aggregate agricultural production and employment would need to be further investigated, if undesirable social effects are to be avoided.

A number of alternative or supplementary measures, such as the creation of technical and administrative capabilities to enforce the current restrictive legislation on chemicals, delimitation of environment-sensitive areas and incentives to R&D on environmental friendly technologies, could be implemented. Diffusion of information could be speeded up by joint efforts of governmental and non-governmental organizations. Also the mass media, such as television, which has a major influential role, can be used not only to diffuse information on sustainable practices but also to raise public awareness of environment and health issues.

Technological progress, alternative rural credit policies and premium prices can enhance relative profitability for farmers adopting sustainable practices and can speed up diffusion. Brazil already has some governmental organizations which deal with R&D on sustainable agriculture. Further investment in this area and cooperation with non-governmental organizations could be pursued. The rural credit system would have to be adapted to the economic conditions of smallholdings and the technical/financial requirements of the new practices (e.g. an emphasis on credit resources for payment of labour instead of for purchasing of agrichemicals). Special credit conditions for farmers adopting sustainable practices (a form of cross-compliance) would be one policy option. Government subsidies on prices may be unrealistic in the current Brazilian context, but premium market prices for chemical-free production could be generated if certification schemes were set up. This can be particularly important to Espírito Santo's farmers, given the State is located in a region with a large potential market, but it would require agreement on the range of acceptable practices to qualify for the certification scheme.

Suggestions for future research

This study focuses on adoption and diffusion. The generation of sustainable agricultural technologies was not emphasised here. Sustainable practices have been developed through both traditional approaches to R&D and participatory technology development, which associates R&D with adoption and diffusion. Given the importance of technological progress in determining profitability of sustainable agriculture, future research should further investigate this aspect of the problem.

There are also limitations in terms of data that restrained the methodological approaches of this thesis, and suggest future investigation. First, the data set for duration analysis should ideally comprise information on life history, but this could not be obtained in our case. This limitation constrained the analysis in the sense that other specifications of the baseline hazard could not be tested. Second, the hypothesis that farmers can reverse their decisions suggests a different model of duration analysis. This is a case of more than one change of state (non-adopter/adopter/non-adopter), for which multiple-spell models of duration analysis could be used (Heckman and Singer, 1984). Here, data on farmers who have reversed their decisions would be necessary. Third, a better sustainability index could be developed if more information on farmers' practices were available. This would permit, for instance, a methodological approach in which the effect of explanatory variables on farmers' sustainability (measured by a sustainability index) could be investigated, instead of analysing the effect of those variables in a dichotomous (adoption/non-adoption) setting.

It was shown that the economic context in which farmers operate can play an important part in the decision to adopt sustainable agricultural technologies. Little attention has been given to this point in the literature on sustainable agriculture. The economic environment changes not only over time but also between countries and regions. For instance, the economic attractiveness of sustainable practices in Europe may have changed since the CAP reform. Also, the determinants of adoption in USA and European countries, where agricultural production is still highly subsidized, may differ from those of Brazil. Further investigation of the matter would seem worthwhile.

The process of technological change may result in unforeseen, undesirable consequences in economic and social conditions. Thus, the overall effect of policies stimulating diffusion of environment-friendly technologies should be assessed. One possible option would be to use simulation models to evaluate the effect of such policies, and also the effect of wide diffusion of sustainable practices on input prices, output prices, employment and aggregate production.

Bibliography

Akinola, A.A. and Young, T. (1985), 'An Application of the Tobit Model in the Analysis of Agricultural Innovation Adoption Process: A Study of the Use of Cocoa Spraying Chemicals Among Nigerian Cocoa Farmers', *Oxford Agrarian Studies* Vol. XIV, pp. 26-51.

Akinola, A.A. (1984), *A Theoretical and Empirical Analysis of Innovation Adoption Process: A Case Study of Cocoa Spraying Chemicals in Nigeria*, Ph.D thesis, Department of Agricultural Economics, School of Economic Studies, University of Manchester: Manchester.

Almada, V.P.F. de (1984), *Escravismo e Transição: O Espírito Santo. 1850-1888*, Graal: Rio de Janeiro.

Almeida, J. (1989), 'Propostas Tecnológicas Alternativas na Agricultura', *Cadernos de Difusão Tecnológica*, Vol. 6, No, 2/3, pp. 183-216.

Alternative Agriculture Project Consultants (1994), *Consultants in Agriculture Projects*, Rio de Janeiro.

Anderson, M.D. (1994), 'Economics of Organic and Low-Input Farming in the United States of America' in Lampkin, N.H. and Padel, S. (eds.), *The Economics of Organic Farming: An International Perspective*, CAB International: Wallingford, pp. 161-184.

Anosike, N. and Coughenour, C.M. (1990), 'The Socioeconomic Basis of Farm Enterprise Diversification Decisions', *Rural Sociology*, Vol. 55, No. 1, pp. 1-24.

Arrow, K. (1962), 'The Economic Implications of Learning by Doing', *Review of Economics Studies*, Vol. 29, pp. 155-173.

Associação de Programas em Tecnologia Alternativa (1993), 'Taller Sobre Estrategias para el Desarollo Sostenible en America Latina y el Caribe: Rede de Intercambio e Ajuda Mutua em Agroecologia', unpublished paper, Vitória.

Banco Central, *Boletins do Banco Central*, Banco Central: Brasília.

Beus, C.E. and Dunlap, R.E. (1990), 'Conventional Versus Alternative Agriculture: The Paradigmatic Roots of the Debate', *Rural Sociology*, Vol. 55, No. 4, pp. 590-616.

Beus, C.E. and Dunlap, R.E. (1991), 'Measuring Adherence to Alternative vs Conventional Agricultural Paradigms: A Proposed Scale', *Rural Sociology*, Vol. 56, No. 3, pp. 432-460.

Biggs, S. (1990), 'A Multiple Source of Innovation Model of Agricultural Research and Technology Promotion', *World Development*, pp. 1481-1499.

Bishop, R. (1978) 'Endangered Species and Uncertainty: The Economics of a Safe Minimum Standard', *American Journal of Agricultural Economics*, Vol. 60, No. 1, pp. 11-18.

Bonanno, A. (1991), 'From an Agrarian to an Environmental, Food, and Natural Resource Base for Agricultural Policy: Some Reflections on the Case of the EC', *Rural Sociology* 56(4), pp. 549-564.

Bonus, H. (1973), 'Quasi-Engel Curves, Diffusion, and the Ownership of Major Consumer Durables', *Journal of Political Economy*, 81 (1973), pp. 655-677.

Buttel, F. H. (1992), 'Environmentalization: Origins, Process, and Implications for Rural Social Change', *Rural Sociology*, 57(1), pp. 1-27.

Buttel, F.H., Gillespie Jr., G.W., Larson III, O.W., and Harris, C.K. (1981), 'The Social Bases of Agrarian Environmentalism: A Comparative Analysis of New York and Michigan Farm Operators', *Rural Sociology*, 46(3), pp. 391-410.

Cano, W. (1985), *Desequilíbrios Regionais e Concentração Industrial no Brasil: 1930-1970*, Global: São Paulo.

Central Bank of Brazil (1992), *Brazil Economic Program*, No. 35, Brasília.

Collet, D. (1994), *Modelling Survival Data in Medical Research*, Chapman & Hall: London.

Colman, D., Crabtree, B., Froud, J., O'Carroll, L. (1992), *Comparative Effectiveness of Conservation Mechanisms*, Department of Agricultural Economics, Faculty of Economic and Social Studies, University of Manchester, Manchester.

Comissão Coordenadora do Relatório Estadual sobre Meio Ambiente e Desenvolvimento (1991), *Meio Ambiente e Desenvolvimento no Espírito Santo*, Vitória.

Conway, G.R. (1987), 'The Properties of Agrosystems', *Agricultural Systems*, No. 24, pp 95-117.

Conway, G.R. and Barbier, E.B. (1990), *After the Green Revolution: Sustainable Agriculture for Development*, Earthscan: London.

Coombs, R., Saviotti, P., Walsh, V. (1987), *Economics and Technological Change*, Macmillan: London.

Cox, D.R. (1975), 'Partial Likelihood', *Biometrika*, 62, 2, pp. 269-276.
Cox, D.R. and Oaks, D. (1984), *Analysis of Survival Data*, Chapman & Hall: London.
Cox, D.R. (1972), 'Regression Models and Life Tables (With Discussion)' *Journal of the Royal Statistical Society* B, 34, pp. 187-220.
Cox, D.R. and Snell, E.J. (1968), 'A General Definition of Residuals', *Journal of the Royal Statistical Society*, A, 30, pp. 248-275.
Cramer, J.S. (1969), *Empirical econometrics*, North Holland: Amsterdam.
D'Souza, G., Cyphers, D., and Phipps (1993), T. 'Factors Affecting the Adoption of Sustainable Agricultural Practices', *Agricultural and Resource Economics Review*, 22 (2), pp. 159-165.
Daberkow, S.G. and Reichelderfer, K.H. (1988), 'Low-Input Agriculture: Trends, Goals, and Prospects for Input Use', *American Journal of Agricultural Economics*, Vol. 70, Part 2, pp. 1159-1166.
Daly, H.E. (1987), 'The Economic Growth Debate: What Some Economists Have Learned But Many Have Not', *Journal of Environmental Economics and Management*, No. 14, pp. 323-336.
Daly, H.E. and Cobb Jr., J.B. (1989), *For the Common Good: Redirecting the Economy Towards Community, the Environment, and a Sustainable Future*, Green Print: London.
David, P.A. (1966), 'The Mechanization of Reaping in the Ante-Bellun Midwest', Chapter 4 in *Technical Choice, Innovation and Economic Growth*, Cambridge University Press: Cambridge, 1975.
Davies, S. (1979), *The Diffusion of Process Innovations*, Cambridge University Press: Cambridge.
de Melo, F.H. (1994), 'Café Brasileiro: Não a Um Novo Acordo Internacional', *Estudos de Política Agrícola*, No. 23, pp. 31-48.
Diebel, P.L., Taylor, D.B., and Batie, S.S. (1993), 'Barriers to Low-Input Agriculture Adoption: A Case Study of Richmond County, Virginia', *American Journal of Alternative Agriculture*, Volume 8, No. 3, pp. 120-127.
Dixon, R. (1980), 'Hybrid Corn Revisited', *Econometrica*, 48, pp. 1451-1461.
Dosi, G. (1991), 'The Research on Innovation Diffusion: An Assessment', in Nakicenovic, N. and Gruebler, A., *Diffusion of technologies and social behavior* (eds), Springer-Verlag: Berlin.
Drandakis, E.M. and Phelps, E.S. (1966), 'A Model of Induced Invention, Growth and Distribution', *Economic Journal*, 76 (1966), 823-840.
EC (1992), 'Council Regulation (EEC) No. 2078/92 of 30 June 1992 on Agricultural Production Methods Compatible With the Requirements of the Protection of the Environment and the Maintenance of the Countryside', *Official Journal of the European Communities*, No. L215/85-90.

ECLAC Economic Commission for Latin America and Caribbean (1991), *Sustainable Development: Changing Production Patterns, Social Equity and the Environment*, United Nations: Santiago.
EMBRAPA Empresa Brasileira de Pesquisa Agropecuaria (1993), *EMBRAPA, Environment & Development*, EMBRAPA-SPI: Brasília.
Espírito Santo (1991), *Diretrizes Para o Setor Agricola 1991-1994*, Secretaria de Estado da Agricultura, Vitoria.
Feder, G. and O'Mara, G.T. (1981), 'Farm Size and the Diffusion of Green Revolution Technology', *Economic Development and Cultural Change*, 30, pp. 50-76.
Feder, G. and O'Mara, G.T. (1982), 'On Information and Innovation Diffusion: A Bayesian Approach', *American Journal of Agricultural Economics*, 64, pp. 145-147.
Feder, G. and Slade, R. (1984), 'The Acquisition of Information and the Adoption of New Technology', *American Journal of Agricultural Economics*, 66, pp. 312-320.
Feder, G., Just, R. and Zilberman D. (1985), 'Adoption of Agricultural Innovations in Developing Countries: A Survey', *Economic Development and Cultural Change*, 33, pp. 255-298.
Flores, M.X., Quirino, T.R., Nascimento, J.C., Rodrigues, G.S. and Buschinelli, C. (1991), *Pesquisa Para Agricultura Auto-Sustentável: Perspectiva de Política e Organização na EMBRAPA*, EMBRAPA-SEA: Brasília.
Fundação Getúlio Vargas (1994), *Conjuntura Econômica*, agôsto, Instituto Brasileiro de Economia, Rio de Janeiro.
Fundação Getúlio Vargas (1995), *Conjuntura Econômica*, agôsto, Instituto Brasileiro de Economia, Rio de Janeiro.
Fundação SOS Mata Atlântica (1993), 'Espírito Santo Perdeu 4.56% de Mata Atlântica Entre 1985 e 1990', *Boletim da Fundação SOS Mata Atlântica*, Fevereiro/93, Ano V, No. 1.
Goldin, I. and Rezende, G.C. de (1990), *Agriculture and Economic Crisis: Lessons from Brazil*, OECD: Paris.
Goldin, I. and Rezende, G.C. de (1993), *Agricultura Brasileira na Década de 80: Crescimento Numa Economia em Crise*, IPEA: Rio de Janeiro.
Goldstein, W.A. and Young, D.L. (1987), 'An Agronomic and Economic Comparison of a Conventional and a Low-Input Cropping System in the Palouse', *American Journal of Alternative Agriculture*, Vol. II, No. 2, pp. 51-56.
Greene, W.H. (1992), *LIMDEP User's Manual and Reference Guide*, Econometric Software, New York.
Greene, W.H. (1993), *Econometric analysis*, Macmillan: New York.

Griliches, Z. (1957), 'Hybrid Corn: An Exploration in the Economics of Technological Change', *Econometrica*, 25, pp. 501-522.
Grubb, M., Koch, M., Munson, A., Sullivan, F. and Thomson, K. (1993), *The Earth Summit Agreements: A Guide and Assessment*, Earthscan: London.
Guarnieri, L.C. (1979), *Alguns Aspectos Sócio-Econômicos do Planejamento na Cafeicultura*, MA dissertation, Universidade Estadual de Campinas, Campinas.
Hannan, T.H. and MacDowell, J.M. (1984), 'The Determinants of Technology Adoption: The Case of the Banking Firm', *Rand Journal of Economics*, Vol. 15, No. 3, pp. 328-335.
Hannan, T.H. and MacDowell, J.M. (1987) 'Rival Precedence and Dynamics of Technology Adoption: An Empirical Analysis', *Economica*, 54, pp. 155-171.
Hanson, J.C., Dale, M.J., Steven, E.P., and Rhonda, R.J. (1990), 'The Profitability of Sustainable Agriculture on a Representative Grain Farm in the Mid-Atlantic Region, 1981-89', *Northeastern Journal of Agricultural and Resource Economics*, 19(2), pp. 90-98.
Harcourt, G.C. (1991), *Some Cambridge Controversies in the Theory of Capital*, Gregg Revivals: Aldershot.
Hartwick, J.M. (1977), 'Intergenerational Equity and the Investing of Rents from Exhaustible Resources', *The American Economic Review*, Vol. 65, No. 5, pp. 972-974.
Hayami, Y. and Ruttan, V. (1973), 'Professor Rosemberg and the Direction of Technological Change: A Comment', *Economic Development and Cultural Change*, 21 (1973), pp. 352-355.
Heckman, J. and Singer, B. (1984), 'Econometric Duration Analysis', *Journal of Econometrics*, 24, pp. 63-132.
Heffernan, W.D. and Green, G.P. (1986), 'Farm Size and Soil Loss: Prospects for a Sustainable Agriculture', *Rural Sociology*, 51(1), pp. 31-42.
Helmar, M.D. (1994) 'Brazilian Agriculture and Policy Changes Under GATT', *GATT Research Paper 94-GATT 2*, Center for Agricultural and Rural Development, Iowa State University.
Henão, S., Nunes, L.A.L., Benatto, A. and Rivera, J.R. (1991), *Diagnóstico Preliminar do Uso de Agrotóxicos no Brasil e Seus Impactos Sobre a Saúde Humana e Ambiental*, Reunião Sobre Agrotóxicos, Saúde Humana e Ambiental no Brasil, mimeo, Brasília.
Henning, J. (1994), 'Economics of Organic Farming in Canada' in Lampkin, N. H. and Padel, S. (eds), *The Economics of Organic Farming: An International Perspective*, CAB International: Wallingford, pp. 143-160.
Hicks, J. R. (1932), *The Theory of Wages*, Macmillan: London.
Hill, L.D. and Kau, P. (1973), 'Application of Multivariate Probit to a Threshold

Model of Grain Dryer Purchasing Decisions, *American Journal of Agricultural Economics*, 55, pp. 19-25.

Hoffmann, R. (1992), 'Vinte Anos de Desigualdade e Pobreza na Agricultura Brasileira', *Revista de Economia e Sociologia Rural*, V. 30, No 2, pp. 97-113, abr/jun, Brasilía.

Hooks, G. M., Napier, T.L. and Carter, M. (1983), 'Correlates of Adoption Behaviors: The Case of Farm Technologies', *Rural Sociology* 48(2), pp. 308-323.

IBGE Instituto Brasileiro de Geografia e Estatística, *Anuários Estatisticos do IBGE*, Fundação Instituto Brasileiro de Geografia e Estatística: Rio de Janeiro.

IBGE (1985), *Censo Agropecuário*, Fundação Instituto Brasileiro de Geografia e Estatística: Rio de Janeiro.

IBGE (1990), *Estatísticas Históricas do Brasil: Séries Econômicas, Demográficas e Sociais*, Fundação Instituto Brasileiro de Geografia e Estatística: Rio de Janeiro.

IBGE (1991), *Censo Demográfico*, Fundação Instituto Brasileiro de Geografia e Estatística: Rio de Janeiro.

Ikerd, J.E. (1993), 'Two Related But Distinctly Different Concepts: Organic Farming and Sustainable Agriculture', *Small Farm Today*, February, pp. 30-31.

Ingco, M.D. (1994), *Agricultural Trade Liberalization in the Uruguay Round: One Step Forward, One Step Back?*, The World Bank: Washington, D. C..

Instituto Biodinâmico de Desenvolvimento Rural (1995), *Diretrizes Para os Padrões de Qualidade Biodinâmico, Demeter, Orgânico*, fifth edition.

Jacob, J. C. and Brinkerhoff, M. B. (1986), 'Alternative Technology and Part-Time, Semi-Subsistence Agriculture: A Survey From the Back-to-the-Land Movement', *Rural Sociology*, 51(1), pp.43-59.

Just, R.E.and Zilberman, D.E. (1983), 'Stochastic Structure, Farm Size, and Technology Adoption in Developing Countries', *Oxford Economic Papers*, 35, pp. 307-328.

Kageyama, A.A. (1986), *Modernização, Produtividade e Emprego na Agricultura: Uma Análise Regional*, Ph.D thesis, Universidade Estadual de Campinas, Campinas.

Kageyama, A.A. and Rehder, P. (1993), 'O Bem-Estar Rural no Brasil na Década de Oitenta', *Revista de Economia e Sociologia Rural*, v. 31, No 1, pp. 23-44. Jan/Mar. Brasilia.

Kalbfleich, J.D. and Prentice, R.L. (1980), *The Statistical Analysis of Failure Time Data*, John Wiley & Sons: New York.

Karshenas, M. and Stoneman, P. (1990), 'Rank, Stock, Order and Epidemic Effects in the Diffusion of New Process Technologies: An Empirical Model', *Warwick Economic Research Papers*, No. 358, April.

Kay, A. (1995), *The Political Economy of the MacSharry Reforms of the CAP*, paper presented at the Agricultural Economics Society Postgraduate Conference at the University of Reading, Reading.

Kelley, M.R. and Brooks, H. (1991), 'External Learning Opportunities and the Diffusion of Process Innovations to Small Firms: the Case of Programmable Automation', in Nakicenovic, N. and Gruebler, A. (eds), *Diffusion of technologies and social behavior*, Springer-Verlag: Berlin.

Kennedy, C. (1964), 'Induced Bias in Innovation and the Theory of Distribution', *Economic Journal*, 74, pp. 541-547.

Kiefer, N.M. (1985), 'Specification Diagnostics Based on Laguerre Alternatives for Econometric Models of Duration', *Journal of Econometrics*, No. 28, pp. 135-154.

Kiefer, N.M. (1988), 'Economic Duration Data and Hazard Functions', *Journal of Economic Literature*, Vol. XXVI, June, pp. 646-679.

Lampkin, N.H. (1994), 'Economics of Organic Farming in Britain' in Lampkin, N. H. and Padel, S. (eds), *The Economics of Organic Farming: An International Perspective*, CAB International: Wallingford, pp. 71-90.

Lampkin, N. H. and Padel, S. (1994), 'Organic Farming and Agricultural Policy in Western Europe: An Overview in Lampkin, N. H. and Padel, S. (eds), *The Economics of Organic Farming: An International Perspective*, CAB International: Wallingford, pp. 437-453.

Lancaster, T. (1979), 'Econometric Methods for the Duration of Unemployment', *Econometrica*, Vol. 47, No. 4, pp. 939-956.

Lancaster, T. (1990), *The Econometric Analysis of Transition Data*, Cambridge University Press: Cambridge.

Lawless, J.F. (1982), *Statistical Models and Methods for Lifetime Data*, John Wiley & Sons: New York.

Levin, S.G., Levin, S.L., and Meisel, J.B. (1987), 'A Dynamic Analysis of the Adoption of a New Technology: The Case of Optical Scanners', *Review of Economics and Statistics*, 86, pp. 12-17.

Lindner, R.K., Fischer, A. and Pardey, P. (1979), 'The Time to Adoption', *Economic Letters*, 2, pp. 187-190.

Lissoni, F. and Metcalfe, J.S. (1993), 'Diffusion of Innovation Ancient and Modern: A Review of the Main Themes', in Rothwell, R. and Dogson, M. (eds), *Handbook of Industrial Innovation*, Elgar: London.

Lockeretz, W., Shearer, G. and Kohl, D.H. (1981), 'Organic Farming in the Corn Belt', in *Science*, vol. 211, pp. 540-546.

Lockeretz, W. (1989), 'Problems in Evaluating the Economics of Ecological Agriculture', *Agriculture, Ecosystems and Environment*, 27, pp. 67-75.

MacRae, J.R., Hill, S.B., Henning, J. and Bentley, A.J. (1990), 'Policies,

Programs, and Regulations to Support the Transition to Sustainable Agriculture in Canada', *American Journal of Alternative Agriculture*, Vol. 5, No. 2, pp. 76-92.

Maddala, G.S. (1977), *Econometrics*, MacGraw-Hill: New York.

Maddala, G.S. (1983), *Limited-Dependent and Qualitative Variables in Econometrics*, Cambridge University Press: London.

Madden, J.P. and Dobbs, T.L. (1990), 'The Role of Economics in Achieving Low-Input Farming Systems', in Edwards, C.A., Lal, R., Madden, O., Miller, R.H., and House, G. (eds), *Sustainable Agriculture Systems*, Soil and Water Conservation Society: Ankeny, Iowa.

Mansfield, E. (1961), 'Technical Change and the Rate of Imitation', *Econometrica*, 2, pp. 741-766.

Mansfield, E. (1968), *Industrial Research and Technological Innovation*, Norton: New York.

Margulis, S. (1988), *The Economy of the Environmental Impact of the Use of Pesticides in Brazil*, Ph.D thesis, Imperial College, Centre for Environmental Technology, London.

Martine, G. (1989), 'Modernização Agrícola e Emprego Rural no Brasil', *Anais do XXVII Congresso Brasileiro de Economia e Sociologia Rural da SOBER*, Piracicaba, pp.162-189.

Marx, K. (1867), *O Capital*, Civilização Brasileira: Rio de Janeiro,1975.

Meadows, D.H., Meadows, D.L., Randers, J. and Behreus, W.W. (1972), *The Limits to Growth: A Report for the Club of Rome's Project on the Predicament of Mankind*, Earth Island: London.

Ministério da Agricultura (1988), *Provárzeas/Profir: Relatório 88*, Brasília.

Monteiro, M.J.C. (Coord.) (1994), *Revisão da Metodologia de Cálculo dos Índices Setoriais Agrícolas - Índice de Preços Pagos Pelos Produtores Rurais (IPP) e Índice de Preços Recebidos pelos Produtores Rurais (IPR)*, IPEA: Brasília.

Mueller, C.C. and Martine, G. (1994), *Modernização Agropecuária e Mudanças na População Rural de Áreas de Agricultura Dinâmica do Brasil: A Década de 1980*. Mimeo.

Napier, L.T., Thraen, C.S., Gore, A. and Goe, W.R. (1984), 'Factors Affecting Adoption of Conventional and Conservation Tillage Practices in Ohio', *Journal of Soil and Water Conservation*, May-June, pp. 205-208.

National Research Council (1989), *Alternative Agriculture*, National Academy Press: Washington, DC.

Nelson, R. (1987), *Understanding Technical Change as an Evolutionary Process*. North-Holland: Amsterdam.

Nelson, R. and Winter, S. (1982), *An Evolutionary Theory of Economic Change*,

Belknap Press of Havard University: Cambridge, Mass..

Norgaard, R.B. (1984), 'Coevolutionary Agricultural Development', *Economic Development and Cultural Change*, pp. 524-546.

Nowak, P. (1987), 'The Adoption of Agricultural Conservation Technologies: Economic and Diffusion Explanations', *Rural Sociology*, 52(2), pp. 208-220.

OECD (1989), *Agricultural and Environmental Policies: Opportunities for Integration*, OECD: Paris.

OECD (1993), *Agricultural and Environmental Policy Integration: Recent Progress and New Directions*, OECD: Paris.

OECD (1994), *Towards Sustainable Agricultural Production: Cleaner Technologies*, OECD: Paris.

Oelhaf, R.C. (1978), *Organic Agriculture*, Osmun & Co: Montclair, Allanheld, New Jersey.

Okun, B. and Richardson, R.W. (1965), *Studies in Economic Development*, Holt, Rinehart and Winston: London.

Padel, S. (1994), *Adoption of Organic Farming as an Example of the Diffusion of an Innovation: A Literature Review on the Conversion to Organic Farming*, Department of Agricultural Sciences, University of Wales, Aberystwyth.

Padel, S. and Lampkin, N.H. (1994), 'Conversion to Organic Farming: An Overview', in Lampkin, N.H. and Padel, S. (eds), *The Economics of Organic Farming: An International Perspective*, CAB International: Wallingford, pp. 295-316.

Pampel, F. and van Es, J.C. (1977), 'Environmental Quality and Issues of Adoption Research', *Rural Sociology*, Vol. 42, No. 1, pp. 55-71.

Paulino, S.R. (1993), *A Indústria de Pesticidas Agrícolas no Brasil: Dinâmica Inovativa e Demandas Ambientais*, M.A. dissertation, Universidade Estadual de Campinas, Campinas.

Pearce, D.W. and Turner, R.K. (1990), *Economics of natural resources and the environment*, Harvester Wheatsheaf: London.

Pearce, D.W., Barbier, E.B. and Markandya, A. (1990), *Sustainable Development: Economics and Environment in the Third World*, Earthscan: London.

Pearce, D., Markandya, A. and Barbier, E.B. (1994), *Blueprint for a Green Economy*, Earthscan: London.

Petersen, T. (1986a), 'Fitting Parametric Survival Models With Time-Dependent Covariates', *Journal of the Royal Statistical Society*, Series C, 35, N0. 3, pp. 281-288.

Petersen, T. (1986b), 'Estimating Fully Parametric Hazard Rate Models With Time-Dependent Covariates', *Sociological Methods & Research*, Vol. 14, February, pp. 219-246.

Pfeffer, J.M. (1992), 'Labor and Production Barriers to the Reduction of Agricultural Chemical Inputs, *Rural Sociology*, 57 (3), pp. 347-362.
Programa de Cooperação em Agroecologia (1992), *Relatório das Experiências em Agroecologia no Rio Grande do Sul*, Convênio SEMAN/PR e PCA/RS, Porto Alegre.
Rahm, M.R. and Huffman, W.E. (1984), 'The Adoption of Reduced Tillage: The Role of Human Capital and Other Variables', *American Journal of Agriculture Economics*, 66, pp. 405-413.
Redclift, M. (1987), *Sustainable Development: Exploring the Contradictions*, Methuend: London.
Reijntjes, C., Bertus, H. and Waters-Bayer, A. (1992), *Farming for the Future: An Introduction to Low-External-Input and Sustainable Agriculture*, Macmillan: London.
Reinganum, J.F. (1981a), 'On the Diffusion of New Technology: A Game Theoretic Approach', *Review of Economic Studies*, XLVIII, 1981, pp. 395-405.
Reinganum, J.F. (1981b), 'Market Structure and the Diffusion of New Technology', *Bell Journal of Economics*, 14, pp. 619-624.
Reis, R. (1987), 'Agrotóxicos: Uma Ameaça de Morte', *Revista Agora*, Abril/87, pp. 23-37.
Robinson, J. (1954), 'The Production Function and the Theory of Capital', *Review of Economic Studies*, XXI, pp. 81-106.
Robinson, J. (1956), *The Accumulation of Capital*, Macmillan: London.
Robinson, J. (1959), 'Accumulation and Production Function', *The Economic Journal*, LXIX, pp. 433-442.
Rocha, H.C. and Cossetti, M.P (1983), *Dinâmica Cafeeira e Constituição de Indústrias no Espírito Santo - 1930-1970*, UFES/CCJE/Dep. de Economia/Núcleo de Estudos e Pesquisas, Vitória.
Rogers, E.M. (1983), *Diffusion of Innovation*, The Free Press: New York.
Rogers, E.M. and Shoemaker, F.F. (1971), *Communication of Innovations*, Collier-MacMillan: London.
Romeiro, A. and Reydon, B.P. (1994), *O Mercado de Terras*, IPEA: Brasilia.
Rosemberg, N. (1969), 'The Direction of Technical Change: Mechanisms and Focusing Devices', *Economic Development and Cultural Change*, No.18.
Russell, N.P. and Fraser, I.M. (1993), *The Potential Impact of Environmental Cross-Compliance on Arable Farming*, mimeo, University of Manchester, Manchester.
Salter, W.E.G. (1966), *Productivity and Technical Change*, Cambridge University Press: Cambridge.
Sandbrook, R. (1992), 'From Stockholm to Rio', *Earth Summit 1992*, Regency

Press: Wickford.

Saviotti, P.P. and Metcalfe, J.S. (1991), 'Present Development and Trends in Evolutionary Economics', in Saviotti, P.P. and Metcalfe, J.S. (eds), *Evolutionary Theories of Economic and Technological Change: Present Status and Future Prospects*, Harwood.

Schmoockler, J. (1966), *Invention and Economic Growth*, Havard University Press: Havard.

Schumpeter, J. (1934), *The Theory of Economic Development*, Harvard University Press: Cambridge, Mass. (reprinted 1974, Oxford University Press).

Schumpeter, J. A. (1939), *Business Cycles: A Theoretical, Historical, and Statistical Analysis of the Capitalist Process*, MacGraw-Hill: London.

Schumpeter, J. (1943), *Capitalism, Socialism, and Democracy*, Harpers & Row: New York.

Shiki, S. (1991), *Agro-Food Policies and Petty Commodity Production in Brazil: Some Implications of Changes in the 1980s*, Ph.D thesis, University College of London, London.

Silveira, J.M.F.J. da (1994), 'Some Notes About Pesticides Industry in Brazil', *Proceedings of the Gottingen Workshop on Pesticide Policies*, Germany.

Silverberg, G., Dosi, G. and Orsenigo, L. (1988), 'Innovation, Diversity and Diffusion: A Self-Organization Model', *The Economic Journal*, 98, pp. 1032-1054.

Smith, S. (1991), 'A Computer Simulation of Economic Growth and Technical Progress in a Multisectoral Economy', in Saviotti, P.P. and Metcalfe, J.S. (eds), *Evolutionary Theories of Economic and Technological Change: Present Status and Future Prospects*, Harwood.

Soares, E. and Vivan, J. (1988), 'Rede de Intercambio: A Experiencia do Espírito Santo', *Proposta*, No. 36, pp. 28-34.

Solow, R.M. (1986), 'On the Intergenerational Allocation of Natural Resources', *Scandinavian Journal of Economics*, No. 88(1), pp. 141-149.

Solow, R.M. (1957), 'Technical Change and the Aggregate Production Function', *Review of Economics and Statistics*, 39, pp. 312-320.

Souza Filho, H.M. de (1990), *A Modernização Violenta: Principais Transformações na Agropecuária Capixaba*, MA dissertation, Universidade Estadual de Campinas, Campinas.

Stineer, D.H., Stinner, B.R. and Paoletti, M.G., (1989) 'In Search of Traditional Farm Wisdom for a More Sustainable Agriculture: A Study of Amish Farming and Society', in Paoletti, M.G., Stinner, B.R. and Lorenzoni, G.G. (eds), *Agricultural Ecology and Environment*, Elsevier: New York.

Stoneman, P. (1980), 'The rate of Imitation, Learning and Profitability', *Economic Letters*, 6, pp. 179-183.

Stoneman, P. (1983), *The Economic Analysis of Technological Change*, Oxford University Press: London.

Taylor, D.C., Mohamed, Z.A., Shamsudin, M.N., Mohayidin, M.G., and Chiew, E.F.C. (1993), 'Creating a Farmer Sustainability Index: A Malaysian Case Study', *American Journal of Alternative Agriculture*, Vol. 8, No. 4, pp. 175-184.

Taylor, D.L. and Miller, W.L. (1978), 'The Adoption Process and Environmental Innovations: A Case Study of a Government Project', *Rural Sociology*, 43(4), pp. 634-648.

Thirtle, C.G. and Ruttan, V.W. (1987), *The Role of Demand and Supply in the Generation and Diffusion of Technical Change*, Harwood Academic Publishers: London.

Thomas, J.K., Ladewig, H. and McIntosh, W.A. (1990), 'The Adoption of Integrated Pest Management Practices Among Texas Cotton Growers', *Rural Sociology*, 5(3), pp.395-410.

Tisdell, C. (1988), 'Sustainable Development: Differing Perspectives of Ecologists and Economists, and the Relevance to LDC', *World Development*, Vol. 16, No. 3, pp. 373-384.

Tremblay, K.R., Jr., and Dunlap, R.A. (1978), 'Rural-Urban Residence and Concern With Environmental Quality: A Replication and Extension', *Rural Sociology*, 43, pp. 474-491.

Usher, A.P. (1954), *A History of Mechanical Inventions*, rev. ed., Harvard University Press: Cambridge, Mass.

Vanclay, F. and Lawrence, G. (1994), 'Farmer Rationality and the Adoption of Environmentally Sound Practices: A Critique of the Assumptions of Traditional Agricultural Extension', *European Journal for Agricultural Education and Extension*, Vol. 1, No. 1, pp. 59-90.

van Liere, K.D. and Dunlap, R.A. (1980), 'The Social Basis of Environmental Concern: A Review of Hypotheses, Explanations and Empirical Evidences', *Public Opinion Quarterly*, pp. 181-197.

Veeman, T.S. (1991), 'Sustainable Agriculture: An Economist Reaction, *Canadian Journal of Agricultural Economics*, Vol. 39, pp. 587-590.

von der Weid, J.M. and Almeida, S.G. de (1988), *Potencialidades y Limitaciones de las Tecnologias Apropriadas para el Desarrollo Agricola en el Contexto de las Actuales Relaciones Entre las ONGs Brasilenas y el Estado*, mimeo, Rio de Janeiro.

von Weizsacker, C.C. (1966) 'Tentative Notes on a Two Sector Model With Induced Technical Progress', *Review of Economic Studies*, 33 (1966), pp. 245-251.

WCED World Commission on Environment and Development (1987), *Our*

Common Future, Oxford University Press: Oxford.

Wernick, S. and Lockeretz, W. (1977), 'Motivations and Practices of Organic Farmers', *Compost Science*, 18(6), pp. 20-24.

Wynen, E. (1994), 'Economics of Organic Farming in Australia', in Lampkin, N.H. and Padel, S. (eds), *The Economics of Organic Farming: An International Perspective*, CAB International: Wallingford, pp. 185-200.

Yomans, K.A. (1968), 'Statistics for the Social Scientists, Vol. 2', *Applied Statistics*, Harmondsworth, Penguin: Middlesex.

Young, T. and Burton, M.P. (1992), 'Agricultural Sustainability: Definition and Implications for Agricultural and Trade Policy', *FAO Economic and Social Development Paper*, No. 110, FAO: Rome.

Appendix

Table A.1
Estimates of exponential hazard functions, adoption of agricultural sustainable technologies, 138 farms

	Equation (1)		*Equation (2)*		*Equation (3)*	
	Estimate	Prob|t|≥X	Estimate	Prob|t|≥X	Estimate	Prob|t|≥X
CONSTANT	-4.738	0.005	-3.099	0.027	-4.884	0.000
SIZE	-0.011	0.031	-0.010	0.041	-0.009	0.070
ACCIDENT	0.878	0.011	0.784	0.021	0.689	0.025
SOCIAL	0.668	0.061	0.760	0.025	0.642	0.045
ENVIRON	0.521	0.118	0.459	0.142	0.395	0.168
F-LABOUR	0.103	0.079	0.126	0.009	0.108	0.019
EXT-NGO	0.945	0.025	0.858	0.017	0.781	0.020
NGO_t	2.381	0.000	2.353	0.000	-	-
$R\text{-}TRADE_t$	-1.138	0.114	-1.198	0.082	-	-
$WAGE\text{-}CHE_t$	-3.943	0.009	-3.925	0.005	-	-
EXT-GO	-0.309	0.426	-	-	-	-
AGE	0.017	0.239	-	-	-	-
RESIDENCE	0.347	0.555	-	-	-	-
EDUCATION	0.061	0.267	-	-	-	-
OWNERSHIP	0.285	0.515	-	-	-	-
OFF-INCOME	-0.000	0.928	-	-	-	-
Log-Likelihood	-207.159		-210.050		-245.358	

Table A.2

Logit and probit models: marginal effects on the probability of adoption of sustainable agricultural technologies, 138 farms *

	Logit				Probit			
	Equation 1		*Equation 2*		*Equation 3*		*Equation 4*	
	Estimate	Prob |t|≥X	Estimate	Prob |t|≥X	Estimate	Prob |t|≥X	Estimate	Prob |t|≥X
CONSTANT	-1.484	0.001	-0.757	0.000	-1.372	0.000	-0.681	0.000
SIZE	-0.571	0.000	-0.005	0.002	-0.005	0.001	-0.004	0.003
ACCIDENT	0.377	0.002	0.322	0.004	0.342	0.002	0.298	0.004
SOCIAL	0.284	0.020	0.278	0.018	0.251	0.026	0.247	0.021
ENVIRON	0.250	0.056	0.248	0.059	0.231	0.064	0.232	0.056
F-LABOUR	0.073	0.007	0.066	0.003	0.058	0.013	0.058	0.003
EXT-NGO	0.482	0.000	0.434	0.001	0.455	0.000	0.399	0.000
EXT-GO	-0.193	0.129	-	-	-0.207	0.079	-	-
AGE	0.006	0.291	-	-	0.007	0.171	-	-
RESIDENCE	-0.009	0.964	-	-	0.012	0.956	-	-
EDUCATION	0.030	0.175	-	-	0.028	0.178	-	-
OWNERSHIP	0.273	0.094	-	-	0.220	0.132	-	-
OFF-INCOME	0.001	0.755	-	-	-0.001	0.695	-	-
Log-Likelihood	-55.485		-60.229		-56.325		-61.203	
Restr.(Slopes=0)Log-L	-94.943		-94.943		-94.943		-94.943	
Correct predictions, adopters	85.5%		82.3%		85.5%		82.3%	
Correct predictions, n-adopt.	81.6%		88.2%		84.2%		88.2%	

* Marginal effects were calculated at mean values of regressors

Table A.3
Explanatory variables correlation matrix, 141 observations

	SIZE	ACCIDENT	SOCIAL	ENVIRON	F-LABOUR	EXT-NGO	EXT-GO	AGE	RESIDENCE	EDUCATION	OWNERSHIP	OFF-INCOME	NGO_t	R-TRADEt	$WAGE\text{-}CHE_t$
SIZE	1.00	-0.03	-0.06	-0.04	0.19	-017	0.03	0.25	0.13	-0.05	0.14	-0.16			
ACCIDENT		1.00	0.17	0.03	0.07	0.27	-0.12	-0.10	0.22	-0.06	-0.04	-0.05			
SOCIAL			1.00	0.04	-0.02	0.32	0.01	-0.14	0.04	0.11	0.05	0.07			
ENVIRON				1.00	0.05	0.14	-0.02	-0.13	0.05	-0.01	0.10	0.04			
F-LABOUR					1.00	-0.02	-0.17	0.49	-0.32	-0.31	-0.09	-0.23			
EXT-NGO						1.00	-0.10	-0.34	0.14	0.01	0.04	0.02			
EXT-GO							1.00	-0.02	-0.03	-0.08	-0.06	-0.04			
AGE								1.00	0.18	-0.41	-0.02	-0.12			
RESIDENCE									1.00	-0.28	0.08	-0.33			
EDUCATION										1.00	0.09	0.28			
OWNERSHIP											1.00	0.01			
OFF-INCOME												1.00			
NGO_t													1.00	0.02	0.27
$R\text{-}TRADE_t$														1.00	-0.05
$WAGE\text{-}CHE_t$															1.00

For Product Safety Concerns and Information please contact our EU representative GPSR@taylorandfrancis.com Taylor & Francis Verlag GmbH, Kaufingerstraße 24, 80331 München, Germany

T - #0137 - 270225 - C0 - 212/152/10 - PB - 9781138384446 - Gloss Lamination